Regulating from the Inside

Can Environmental Management Systems Achieve Policy Goals?

edited by
Cary Coglianese and Jennifer Nash

RESOURCES FOR THE FUTURE $=$ WASHINGTON, DC

Printed in the United States of America

An RFF Press book
Published by Resources for the Future
1616 P Street, NW, Washington, DC 20036–1400
www.rff.org

Library of Congress Cataloging-in-Publication Data

Regulating from the inside : can environmental management systems achieve policy
    goals? / edited by Cary Coglianese and Jennifer Nash.
        p.   cm.
    Includes bibliographical references and index.
    ISBN 1–891853–40–6 (lib. bdg.) — ISBN 1–891853–41–4 (pbk.)
        1. Environmental management. 2. Environmental policy. I. Coglianese, Cary. II.
Nash, Jennifer, 1956–
GE300 .R44   2001
363.7'05—dc21

                                                                    2001019431

f e d c b a

The paper in this book meets the guidelines for permanence and durability of the Committee on Production Guidelines for Book Longevity of the Council on Library Resources.

The text of this book was designed and typeset in Trump Medieval and ITC Franklin Gothic by Betsy Kulamer. It was copyedited by Pamela Angulo. The cover was designed by Debra Naylor Design.

ISBN 1–891853–40–6 (cloth)   ISBN 1–891853–41–4 (paper)

About
Resources for the Future
and *RFF Press*

Founded in 1952, **Resources for the Future** (RFF) contributes to environmental and natural resource policymaking worldwide by performing independent social science research.

RFF pioneered the application of economics as a tool to develop more effective policy about the use and conservation of natural resources. Its scholars continue to employ social science methods to analyze critical issues concerning pollution control, energy policy, land and water use, hazardous waste, climate change, biodiversity, and the environmental challenges of developing countries.

RFF Press supports the mission of RFF by publishing book-length works that present a broad range of approaches to the study of natural resources and the environment. Its authors and editors include RFF staff, researchers from the larger academic and policy communities, and journalists. Audiences for RFF publications include all of the participants in the policymaking process—scholars, the media, advocacy groups, NGOs, professionals in business and government, and the general public.

Contents

Foreword

This book stands squarely at the crossroads of contemporary environmental policy. Within the last generation, many countries around the world have made significant strides in improving environmental conditions. In the United States, for example, we have come a long way. Cleveland's Cuyahoga River, which once caught fire, now features cruise boats. The Los Angeles skyline, once invisible, now can be readily seen, at least on good days. Local communities are doing a much better job managing their landfills, and fish are returning to the Hudson River.

Nevertheless, we clearly have a very long way to go. On a recent plane flight, a friend of mine described one piece of the problem. "I represent big turkey farms," he said, "and like it or not, that's the future of turkey production. We all have romantic notions of small breeders tending to small flocks, but the large farms will dominate the future." These large, sprawling complexes create new kinds of pollution problems. The turkeys create huge quantities of waste that must be managed responsibly. If the turkey farmers do not stay ahead of the waste, heavy rains can wash it into the watershed, where it causes tough and complicated problems downstream. The farms draw a lot of water from streams and wells, and that causes a different kind of watershed problem. And to the people who live nearby, the waste of thousands of turkeys presents an obvious air pollution issue.

Other "new-generation" problems are lurking as well. Small dry-cleaning establishments have noxious chemicals needing disposal. Growing piles of obsolete computers create new hazardous waste problems. Fertilizer runoff from farms challenges many ecosystems, and managing storm water runoff is a fast-growing problem in burgeoning urban areas. Meanwhile, a large array of nasty, unresolved "old-generation" problems still challenges the system.

What should we do? Many environmental groups argue strongly that the problem is not that we need new strategies; we just need to follow through on the old ones more vigorously. These groups see talk about new approaches to environmental policy as an effort to backtrack from the laws already on the books—which have worked and would work even better if the federal and state governments appropriated enough money to enforce them. Of course, these advocates have good cause for their worries. Past "regulatory relief" strategies, in fact, sometimes have been thinly veiled efforts to roll back environmental standards.

Without a doubt, more aggressive enforcement of existing environmental standards could bring even more progress. However, there is doubt about whether the advocates of this approach can assemble enough political support. The Republicans' abrupt failure to shrink the U.S. Environmental Protection Agency (EPA) during the "Contract with America" days, following their takeover of Congress in 1995, demonstrated that Americans care a great deal about protecting the environment. In the years afterward, though, it proved difficult to get a substantial increase in resources for EPA. Legislators eyed tax cuts and prescription drugs, rather than more money for environmental regulation, as destinations for the budget surplus.

Moreover, even though skillful government managers could tinker with their regulatory tools to make them fit new-generation problems, such as nonpoint-source pollution, doing so would be like using an expensive screwdriver to pry open a paint can: it would work, but other tools could do the job better and cheaper. In addition, as some of the authors in this book point out, the choice does not have to be either/or. Regulators can continue to use the old-generation tools for old-generation problems. They also can develop new-generation tools for problems that simply don't respond to the old tools. They can use old-generation sticks for problems where enforcement works best, and they can develop new-generation carrots for problems—and players— where innovation offers a significant breakthrough.

This book, in short, is about whether—and how—government policymakers can develop a more sophisticated environmental toolbox. It is a search for new ways to make the environment cleaner and healthier without sacrificing hard-won goals. It is also a search for new collaborative tools for government regulators and the entities they regulate. The basic questions include:

■ Can private-sector organizations devise environmental management systems to ensure corporatewide compliance with environmental laws?
■ Can the private sector use these systems to go beyond compliance with existing regulations?

▓ Can government develop different tracks for its environmental oversight to accommodate such innovative strategies? At the same time, can it manage a more traditional regulatory track for less-innovative—or low-performing—organizations?

▓ Can government devise a system that citizens can understand and trust?

These questions focus sharply some difficult and fascinating problems. For a generation, the pursuit of a clean and safe environment has been based on applying tough standards to all. Even if the regulations are tough, there always has been some comfort in knowing that they applied equally to every-one. Different regulatory strategies for different kinds of organizations might yield more environmental improvement at much lower costs. However, rais-ing this issue fuels concerns that the new strategies are a subtle rollback of environmental standards—a secret way of using "regulatory relief" once again to benefit some well-placed corporate interests.

The efforts described in this book are in their infancy, but they are intrigu-ing and promising. The chapters make a strong case that environmental management systems should be analyzed as potential options for solving some of the new-generation environmental policy problems and for produc-ing larger gains for old-generation problems—and with lower costs across the board. Lurking beneath this strategy is a whole new set of interesting policy, political, and ethical questions; this strategy is nothing less than a renegotia-tion of the basic environmental compact between citizens and government. Is it possible, as one of the authors in this book asks, to "get the deal right"? Can a system built on providing more flexibility but holding companies accountable for results produce a cleaner, safer environment while spending less money? Will the public trust a performance-based system? Will the pub-lic see it as an improvement over command-and-control regulation?

In many ways, these questions are central not only to twenty-first–century environmental policy, but also to twenty-first–century governance. This book is really about the strategic use of information as a governance tool—and the possibility of substituting information for traditional govern-mental authority. Environmental management systems are about using information to focus on results and about managing for results instead of managing by regulatory process. Government is wrestling with these puzzles in other policy areas as well, so the findings presented in this book have great potential for informing a much broader debate about the role of govern-ment in the information age. That makes this book important reading, not only for insights into the future of environmental policy but also for a glimpse into the broader future of governance as well.

DONALD F. KETTL
Robert M. La Follette School of Public Affairs
University of Wisconsin–Madison

Acknowledgements

The origins of this book trace back to 1996, when state governments formed the Multi-State Working Group (MSWG) on Environmental Management Systems and invited leaders from business, the U.S. Environmental Protection Agency (EPA), universities, and the public interest community to join the states in studying innovations in environmental management. A couple of years after MSWG was created, its founders had the vision to launch a series of research roundtables at six universities across the United States, one of which was held at Harvard University in August 1998. Shortly afterward, MSWG organized a team of researchers and practitioners who met in Washington, DC, and then again at Harvard to design a research agenda around environmental management systems (EMSs) and their potential for achieving public policy goals.

This book is the result of these efforts to encourage systematic reflection about EMSs and their policy implications. Initial drafts of the chapters provided the basis for an exceedingly productive dialogue in Washington, DC, at a meeting cosponsored by MSWG and the Brookings Institution, the Council of State Governments, the National Academy of Public Administration, and EPA in November 1999. Additional support for this meeting was provided by the American Chemistry Council, the Global Environmental Reporting Initiative, Steptoe & Johnson LLP, and the states of California, Illinois, and Wisconsin.

Several individuals and organizations deserve special acknowledgement for helping to launch and sustain the efforts that led to this book. Robert Stephens, Jeff Smoller, and Peter Wise (from the states of California, Wisconsin, and Illinois, respectively) reached out early to the academic community, believing that university-based researchers have a special contribution to make in understanding the role of EMSs in public and private innovation. Jim Horne, director of EPA's Office of Water, saw the value of EMS research early on, and his office has supported the development of the National Database on Environmental Management Systems. The Ash Fund for Research on Democratic Governance at Harvard University's John F. Kennedy School of Government, EPA's Emerging Strategies Division, and the Visions of Governance for the 21st Century Project supported our efforts to edit this book. MSWG ensured that we had the necessary administrative support.

In addition to those who contributed chapters to this book, we are indebted to the following people who generously offered ideas and energy: Jay Benforado, Sara Burr, Scott Butner, Walt Carey, Marian Chertow, Terry Davies, Suzanne Maria Dickerson, John Ganzi, Dave Guest, John Harris, David Lazer, Henry Lee, John Linglebach, Keri Luly, Rick Minard, David Monsma, Kimberly Nelson, Chris Paterson, Ed Quevedo, Christian Richter, Dan Sprague, Peter Winsemius, and Gayle Woodside. DeWitt John produced a thoughtful synthesis of our Washington, DC, dialogue that helped us as we reflected on the themes of this project. The views expressed in this book do not, of course, necessarily represent the opinions of the many individuals and organizations who lent their support to this effort.

We are grateful to Alice Andre-Clark for research assistance, to Camilia-kumari Wankaner for her exceptional efficiency in preparing the manuscript, and to Carla De Ford for proofreading. Finally, we offer our thanks to Don Reisman at Resources for the Future/RFF Press for the trust he put in this undertaking, to our reviewers for helping us seek continuous improvement in these pages, and to Gina Armento, Pamela Angulo, and Betsy Kulamer at RFF Press for supporting the "plan–do–check–act" process that brought this book to completion in a timely manner.

CARY COGLIANESE
JENNIFER NASH
John F. Kennedy School of Government
Harvard University

Contributors

Deborah Amaral is an adjunct professor of practice in the Curriculum in Public Policy Analysis at the University of North Carolina (UNC) at Chapel Hill and UNC project manager for the National Database on Environmental Management Systems. Previously, she was a senior decision analyst at Lumina Decision Systems and a senior chemical engineer and scientist with ECR, Inc., and held positions in environmental research and teaching.

Richard N.L. (Pete) Andrews is a professor of environmental policy in the Curriculum in Public Policy Analysis; the Carolina Environmental Program; the Kenan Institute of Private Enterprise; and the Department of Environmental Sciences and Engineering, School of Public Health, all at the University of North Carolina (UNC) at Chapel Hill. He researches and teaches about environmental policy in the United States and worldwide; has written a book on the history of U.S. environmental policy; and has received grants to study environmental policy innovations in the United States, the Czech Republic, and Thailand. Beyond the university, he is currently chairing a National Academy of Public Administration study panel on ISO 14001 auditing and registration practices for environmental management systems in the United States. He directs the National Database on Environmental Management Systems.

Cary Coglianese is an associate professor of public policy at Harvard University's John F. Kennedy School of Government and chair of the Regulatory Policy Program at the school's Center for Business and Government. His interdisciplinary research focuses on issues of regulation and administrative law, with a particular emphasis on the empirical study of regulatory reform initiatives. He has published widely on the effectiveness of alternative means of designing regulatory processes (such as negotiated rulemaking) as well as issues related to democratic governance, federalism, international regulation, and environmental policy. A lawyer and political scientist, he also is an affiliated scholar at the Harvard Law School and the director of the Kennedy School's Politics Research Group.

Nicole Darnall is a doctoral student in the Curriculum in Public Policy Analysis at the University of North Carolina at Chapel Hill and a research associate for the National Database on Environmental Management Systems. She previously held research positions with Resources for the Future and the U.S. Department of Agriculture Forest Service.

Derek Davison is a research associate at the Heinz School of Public Policy and Management, Carnegie Mellon University. His research examines the community benefits of environmentally conscious manufacturing techniques and the ways that cities can use lifestyle amenities to attract high-technology workers and companies. He received his masters of science degree in public policy and management from Carnegie Mellon University in 1998.

John R. Ehrenfeld is visiting scholar at the Delft University of Technology. Previously he was a senior research associate in the Center for Technology, Policy, and Industrial Development at the Massachusetts Institute of Technology, with additional appointments as lecturer in the Department of Chemical Engineering and the Department of Urban Studies and Planning, as well as directing the MIT Technology, Business and Environment Program. Prior to his appointments at MIT, he spent more than 20 years in the field of environmental policy and management.

Eric Feldman, formerly a senior research associate at the Environmental Law Institute, is currently a candidate for a masters degree in city planning at the Massachusetts Institute of Technology.

Richard Florida is the H. John Heinz III Professor of Regional Economic Development at the Heinz School of Public Policy and Management, Carnegie Mellon University. He has been a visiting professor at the Massachusetts Institute of Technology and at Harvard University's John F. Kennedy School of Government. His books include *Industrializing Knowledge* (with Lewis Branscomb and Fumio Kodama), *Beyond Mass Production* (with Martin Kenney), and *The Breakthrough Illusion* (with Martin Kenney). He has served as an adviser to the White House Office of Science and Technology Policy, the U.S. Department of Commerce, the U.S. Congress, state and local governments, the Canadian government, the European Union, the Japanese government, and multinational corporations.

Deborah Rigling Gallagher is a doctoral student in the Curriculum in Public Policy Analysis at the University of North Carolina at Chapel Hill and a research associate for the National Database on Environmental Management Systems. Previously, she was an environment, health, and safety manager for Kraft Foods.

Jessica D. Jacoby, formerly a senior research associate at the Environmental Law Institute, is a student at the University of Virginia School of Law.

Suellen Terrill Keiner directs the Center for the Economy and the Environment at the National Academy of Public Administration and manages the National Database on Environmental Management Systems on behalf of the Environmental Law Institute, where she formerly directed the Program on Environmental Governance and Management.

Matthew L. Mitchell is a senior research associate at the Environmental Law Institute.

Shelley H. Metzenbaum is a visiting professor at the University of Maryland's School of Public Affairs, where she runs the Environmental Compliance Consortium, a collaborative effort by states to improve the measurement and management of environmental compliance and enforcement programs. She also runs the Public Sector Performance Management Project at Harvard University's John F. Kennedy School of Government. Previously, she served as associate administrator for regional operations and state–local relations for the U.S. Environmental Protection Agency and as undersecretary of the Massachusetts Executive Office of Environmental Affairs.

William R. Moomaw is a professor of international environmental policy and director of the International Environment and Resource Policy Program at the Fletcher School of Law and Diplomacy, Tufts University. He also serves as director of the Tufts Institute of the Environment. His disciplinary training in chemistry has shaped his role as a translator of science-based issues such as climate change, natural resource use, and chemical contamination into policy-relevant terms.

Jennifer Nash is director of the Regulatory Policy Program at Harvard University's John F. Kennedy School of Government. Her research has investigated the reasons why firms adopt environmental practices that go beyond regulatory requirements. She has published articles in *California Management Review; Business, Strategy and the Environment;* the *Annual Review of Energy and the Environment;* and *Environment* magazine. Previously, she was associate director of the Technology, Business, and Environment Program at the Massachusetts Institute of Technology and executive director of the Delaware Valley Citizens' Council for Clean Air.

Theodore Panayotou is director of the Environment and Sustainable Development Program at the Center for International Development at Harvard University and the John Sawhill Lecturer in Environmental Policy at Harvard's John F. Kennedy School of Government. He specializes in natural resource management and environmental economics as they relate to economic development. An international consultant and economic advisor, he has written *Green Markets* (1993) and *Instruments of Change* (1998) and is the editor of *Environment for Growth* (2001).

Jerry Speir is director of the Institute for Environmental Law and Policy and adjunct professor of law at Tulane University Law School. He is a member of the U.S. Technical Advisory Group to Technical Committee 207 of the International Organization for Standardization and of the Multi-State Working Group on Environmental Management Systems. He recently completed a research project on state innovation programs, using environmental management systems for the National Academy of Public Administration's project entitled *environment.gov: Transforming Environmental Protection for the 21st Century*. He also co-authored *Managing a Better Environment: Opportunities and Obstacles for ISO 14001 in Public Policy and Commerce*.

Regulating from the Inside

Environmental Management Systems and the New Policy Agenda

Cary Coglianese and Jennifer Nash

The past three decades of environmental regulation have resulted in considerable improvements in some of the more visible and pressing environmental impacts of industrial activity. Nevertheless, substantial challenges remain ahead in providing still more effective and sensible environmental protection. To meet the remaining challenges, policymakers and analysts are increasingly looking for a "third way" of dealing with environmental problems, that is, for new policy tools that fall somewhere between the free market and conventional public regulation (Ayres and Braithwaite 1992; The Aspen Institute 1996; Chertow and Esty 1997). In this book, we examine the potential of one such tool that is emerging on the policy agenda: the environmental management system (EMS).

What Is an EMS?

Unlike public regulation, which imposes requirements on organizations from the outside, an EMS consists of a regulatory structure that arises from within an organization. An EMS represents a collection of internal efforts at policymaking, planning, and implementation that yields benefits for the organization as well as potential benefits for society at large (Orts 1995; Fiorino 1999). EMSs appear to many to promise improvement in solving environmental problems. When people inside an organization take responsibility for managing environmental improvement, the internal regulatory strategies they adopt will presumably turn out to be less costly and perhaps even more effective than they would be under government-imposed standards. Moreover, when organizations have the flexibility to create their own internal regulatory approaches, they are more likely to innovate and will potentially find solutions that government standard-setters would never have considered. Finally, individuals within organizations may be more likely to see their organization's own standards as more reasonable and legitimate, which may in turn enhance compliance with socially desirable norms (Kleindorfer 1999).

Business organizations are developing EMSs in increasing numbers. Thousands of European and Asian firms have adopted EMSs in recent years, and an ever-growing number of corporations in the United States are developing their own EMSs and certifying that they meet international EMS standards. EMSs set forth internal rules, create organizational structures, and direct resources that managers use to routinize behavior in order to help satisfy their organizations' environmental goals. Although the specific institutional features of EMSs vary across organizations, under almost any system managers establish an environmental policy or plan; implement the resulting plan by assigning responsibility, providing resources, and training workers; check progress through systematic auditing; and act to correct problems. In some cases, the organization's EMS process draws in outsiders, with firms involving interested members of the community in their environmental planning and relying on independent environmental auditors to help monitor and certify their environmental performance.

The experience of the Louisiana-Pacific Corporation (LP) illustrates the kind of EMS that is the focus of this book. Headquartered in Portland, Oregon, LP is one of the largest manufacturers of building products in North America, with nearly $3 billion in annual sales and more than 80 manufacturing facilities. Like other companies of its size, the manufacturing processes at LP's facilities have a broad range of effects on the environment, from consuming resources such as water and energy to generating solid waste for disposal and emitting pollutants into the air and water.

In the early 1990s, the U.S. Environmental Protection Agency (EPA) filed a suit against LP for unlawful releases of volatile organic compounds. EPA suspected that managers at one of LP's facilities had tampered with air pollution controls, and the firm became the target of a criminal investigation. Realizing the potential for liability, LP hired a corporate environmental manager to bring all the firm's facilities into compliance. The new manager, Elizabeth Smith, quickly realized the challenge she faced: to ensure compliance at all the manufacturing sites scattered across North America. Smith found that she needed a corporate governance structure that could drive environmental responsibilities down into the plants. What she needed, she discovered, was an EMS.

Smith began by identifying the company's regulatory responsibilities and worked to establish an environmental manager position at each plant. She then created a reporting structure for environmental compliance that began within each plant and worked its way up to the company's board of directors. These steps established an institutional structure for environmental management within the corporation, with dedicated lines of responsibility and authority.

At each plant, Smith and her environmental managers organized teams of workers (for example, press operators, drier operators, and maintenance workers) for the purpose of developing standard operating procedures that would be incorporated into the company's EMS. With the assistance of environmental experts, the teams reviewed their plant's existing waste, air, and water permits and then identified the different job functions that were key to ensuring that these permit requirements would be met. For example, drier operators needed to be sure that certain temperatures were maintained, and maintenance workers needed to check key control equipment each hour. Workers with key roles in compliance then wrote standard operating procedures for their jobs, and the corporate staff established extensive training programs to ensure that all employees were informed of these job tasks. Furthermore, the plant teams developed "consequence programs," ensuring that the standard operating procedures would have bite. The result: any worker who violates standard operating procedures faces disciplinary action. In addition, the corporation has put in place a self-inspection program to monitor the effectiveness of its procedures.

In creating a new set of standard operating procedures, LP's EMS established a new set of routines for how workers did their jobs. Workers who in the past thought of themselves as maintaining the cleanliness of the plant came to think of themselves as responsible for the cleanliness of the air, too. The EMS created a structure whereby people came together and talked about regulatory compliance. LP also implemented a process for regular review of and updates to standard operating procedures, to ensure ongoing focus on environmental improvement. LP reports that, over time, workers have begun to suggest work routines that are not strictly based on achieving compliance. For example, new standard operating procedures at LP's plant in Hines, Oregon, now provide for using wood planer shavings in the manufacture of fiberboard products. Planer shavings from this facility had previously been disposed of at a cost to the company, but now they earn the company revenues as inputs to saleable products.

The story of LP's experience in developing an EMS can be played out for many other firms (for example, Balta and Woodside 1999; Carter 1999; Furrer and Hugenschmidt 1999). Overall, the experience of firms with EMSs reveals a common pattern of planning, monitoring, and acting to achieve environmental improvement. This experience also exhibits great variation in the ambitiousness and design of EMSs within firms. We would not want to suggest that the EMS as implemented at LP is the norm for all firms with EMSs. Indeed, it is often thought that one of the virtues of EMSs over traditional regulation is that EMSs are adaptable to the organizations that create and use them. This adaptability offers the potential of permitting more cost-effective and innovative solutions to environmental problems. That same adaptability, of course, poses challenges to researchers seeking to generalize about the impact of EMSs on the attainment of public policy goals.

Not All EMSs Are the Same

EMSs vary not only because they arise in different organizational settings but also because firms adhere to different external standards for EMSs—and in some cases, to no external standards at all, because such standards are voluntary. We should distinguish, therefore, an EMS that is created by managers of a particular firm or facility from the external EMS standards to which that firm's EMS adheres.

EMS standards or guiding principles have emerged from various sources. In the United States, trade associations have developed EMS standards for use by their members. Leading examples are the Responsible Care program of the American Chemistry Council[1] (Nash and Ehrenfeld 1997; Rees 1997) and the Sustainable Forestry Initiative of the American Forest and Paper Association (Meidinger 1999–2000). A handful of other trade associations in the textiles, petroleum, and chemical distribution industries also have developed EMS guidelines. In addition, firms affiliated with the International Chamber of Commerce have created the *Business Charter for Sustainable Development,* which contains 16 principles that thousands of firms have used as a basis for their EMSs. The Coalition for Environmentally Responsible Economies, a group of environmental and investment organizations, has also offered 10 sustainability-based principles that nearly 100 organizations are using to structure their environmental management programs. Most recently, the North American Commission for Environmental Cooperation (the multinational environmental authority established under the North American Free Trade Agreement [NAFTA]) issued a guidance document detailing 10 principles an effective EMS should strive to follow.

In addition, beginning in the early 1990s, standards organizations in Great Britain, Ireland, France, and Spain developed national EMS guidelines. Around the same time, the European Union began developing its own EMS standard, known as the Eco-Management and Audit Scheme (EMAS). The plethora of conflicting national standards led to the ISO 14001 principles adopted by the International Organization for Standardization (ISO) in 1996. More than 15,000 organizations—and the number keeps growing—have formally registered their EMSs as adhering to the ISO 14001 standard, and thousands of other organizations have implemented ISO 14001 systems without having registered them.

The EMS standards developed by trade associations, standards organizations, and others vary along five key dimensions:

- the ambitiousness of the environmental objectives they require managers to establish,
- the trustworthiness of the EMSs they specify,
- the level of monitoring they call for,
- the type of sanctions they impose on firms that do not measure up, and
- the transparency of the EMS and of the organization's performance to the public.

The future of environmental policy should be guided not only by research about whether EMSs achieve socially desirable outcomes—and under what conditions—but also by an understanding of how differences in key characteristics of EMSs affect organizational performance.

First, EMSs that follow different standards will differ in terms of the stringency or ambitiousness of their environmental goals. The Responsible Care program, for example, calls on firms to adopt relatively ambitious EMSs that move firms toward "no accidents, injuries, or harm to the environment" (American Chemistry Council 2000). The American Forest and Paper Association's Sustainable Forestry Initiative calls on firms to meet the needs of the present without compromising future generations, a challenge for any organization (AF&PA 2000). In comparison, the ISO 14001 standard is less ambitious; the substantive requirements it currently imposes are that a firm's internal environmental policy include the goal of regulatory compliance and that the firm explicitly commit itself to improving its EMS.

Second, EMSs that follow different standards will vary in terms of the degree to which they seek to ensure their own credibility and consistency. Some trade association EMS guidelines, for example, require only that managers establish an environmental policy and declare their commitment to achieving it. ISO 14001, in contrast, requires consistency between what managers say they will do and what they actually practice. If an EMS has been credibly certified as meeting the ISO 14001 standards, then people outside the firm have reason to trust that actual behavior corresponds to procedures documented in the EMS.[2]

Third, EMSs implemented in accordance with differing standards will likely vary in the type and extent of monitoring they use. ISO 14001 requires that managers monitor their progress at regular intervals and offers its own system for external verification by accredited third parties. In contrast, the American Textile Manufacturers Institute requires only that firms subscribing to its "environmental excellence" principles describe progress to the trade association annually. This difference holds direct implications for public policy, because monitoring by credible third parties may need to be a prerequisite before accepting an EMS as a substitute for conventional regulatory strategies.

Fourth, EMSs may be implemented with varying degrees of precision because of differences in the severity of sanctions for firms that fail to adhere to different EMS standards. Several trade associations, including the American Forest and Paper Association, have sanctioned their members with expulsion for failing to adopt their EMS specification. Other trade associations are more subtle about imposing sanctioning authority. Anecdotal evidence indicates that some firms have lost their ISO 14001 registration as a result of repeated lapses. The likely impact that loss of certification or trade association membership has on any given firm will itself be quite variable and difficult to predict, but the issue of sanctions will undoubtedly be relevant to policy decisionmaking.

Finally, EMSs developed to conform with different standards provide different levels of information to the public about a firm's EMS and its environmental performance. The only environmental information a firm that has adopted ISO 14001 must publicly disclose is its environmental policy. U.S. firms that adopt Responsible Care must establish a policy of openness with surrounding communities but decide for themselves which information to disclose about their operations. About half of the American Chemistry Council's members have had their Responsible Care programs externally verified, but results of these assessments generally are not available to the public. Firms participating in the Canadian Responsible Care program, however, must provide a copy of their verification report to anyone who requests it.

The variation in the goals, credibility, monitoring, sanctions, and transparency found in different EMS guidelines suggests a lesson for those engaged in research and policymaking. Generalizations about unspecified EMSs probably will prove misleading, mistaken, or both. Moreover, key differences can exist even between systems implemented under the same external EMS standard, because not all firms implementing EMSs that meet such standards will do so with the same level of commitment. External standards generally require that managers establish environmental targets, but they do not specify the substantive nature of the targets. They require that top managers assign responsibility, but not always who will be responsible. They require training, but not necessarily what workers should learn. Hence, even when EMSs are designed to comply with standards adopted by international organizations or industry associations, the firms' managers usually retain flexibility to develop systems that reflect their own organizations' needs, values, commitments, and resources. Simply put, although EMSs share certain common characteristics, neither all such systems nor all EMS standards are the same.

Regulating from the Outside: Government Regulation and Its Limits

The variability, even malleability, of EMSs contrasts dramatically with the rigidity usually associated with government regulation. Existing environmental regulation consists of extensive government-imposed rules requiring that firms adopt specific technologies or methods designed to protect environmental quality (Portney 2000). Such regulation is justified as needed to correct failures in the market that result in negative environmental impacts on society, but for many observers, the conventional approach of imposing government standards on industry has been only a "crude first-generation strategy" for addressing environmental problems (Ackerman and Stewart 1985).

In the United States, this first generation of environmental protection has spawned a series of major environmental statutes, each with its own corresponding and often voluminous set of federal and state regulations: the Clean Air Act of 1970; the Clean Water Act of 1972; the Safe Drinking Water Act of 1974; the Resource Conservation and Recovery Act (RCRA) of 1976; the Toxic Substances Control Act of 1976; the Comprehensive Environmental Response, Compensation, and Liability Act (CERCLA, or Superfund) of 1980; and all the subsequent amendments to each of these statutes. As the titles of these laws indicate, the current system of environmental protection in the United States is organized largely around the various media—air, water, groundwater, and land—through which pollution travels. Much of the existing regulatory regime depends on technology-based standards and on the permitting of individual facilities by state agencies that operate under approved plans. Furthermore, government agencies have tended to rely on inspections and enforcement actions to ensure compliance, and citizen and environmental groups also have been able to bring their own lawsuits to force compliance.

Three decades of efforts have improved various environmental conditions in the United States (Bok 1996; Davies and Mazurek 1998; Portney and Stavins 2000). National air quality trends indicate significantly reduced levels of the air pollutants targeted under the Clean Air Act. Between 1970 and 1999, the population in the United States grew by 33%, vehicle miles driven rose by 140%, and the gross domestic product increased by nearly 150%; however, at the same time, ambient levels of key air pollutants have declined markedly. Levels of lead in the air have dropped by 94%, carbon monoxide by 57%, sulfur dioxide by 50%, nitrogen oxides by 25%, and ozone by 12% (U.S.

EPA 2000a).[3] Releases of toxic chemicals to air, water, and land declined about 45% during the decade from 1988 (when firms were first required to report toxic release information) to 1998, from nearly 3.4 billion pounds to 1.8 billion pounds (U.S. EPA 2000b). Municipal wastewater treatment plants have substantially reduced the level of human waste in rivers and oceans, and the water quality of many rivers and lakes has improved since the mid-1980s. In addition, the Resource Conservation and Recovery Act has led to significant changes in the ways companies manage their hazardous materials, with land disposal of untreated waste now relatively rare.

Regulation has translated into these improvements in environmental conditions in part by motivating improvements in environmental management. Since at least the 1980s, firms have increasingly accepted their obligation to comply with command-and-control environmental regulations (Fischer and Schot 1993). As illustrated by the case of LP, acceptance of the need to comply—reinforced by the threat of civil and criminal penalties for violation of certain environmental laws—has shaped the environmental management of leading firms (Howes and others 1997). Large firms have established legal and environmental departments staffed with specialists who are responsible for knowing how regulatory requirements apply to their organizations. Firms have established internal compliance auditing functions, thereby assuming some of the control functions of agencies (Johnston 1995), and have implemented pollution prevention strategies to help meet regulatory requirements more efficiently.

Although existing public regulation is credited with inducing dramatic improvements in environmental quality, it also is often criticized for its limitations (Davies and Mazurek 1998). Environmental problems continue to persist, many of which—such as nonpoint water pollution and consumption of natural resources—still fall largely outside regulatory control and probably are not amenable to traditional methods of regulatory control (Elliott and Charnley 1998). Moreover, the standard regulatory approach is often attacked for its inflexibility and costs. Many observers of environmental policy in government and business claim that society has reached a point of diminishing returns for investments in environmental controls, such that further tightening of regulatory standards will yield far less in the way of environmental benefit for each dollar invested (Breyer 1993).

The existing command-and-control approach to regulation, which dictates generic control technologies and processes for firms, is said to be either overinclusive or underinclusive, meaning that uniform standards sometimes require firms to do too much in areas where the costs of regulation exceed the benefits, or too little in areas where the benefits of regulation would outweigh the costs (Graham and Wiener 1995; Hahn 1996). Such regulations also fail to account for differences in firms' marginal cost functions, thus raising the concern that similar environmental outcomes could be achieved at lower cost (Pildes and Sunstein 1995). In addition, by establishing rigid requirements for the use of specified technologies, existing regulation may discourage the innovation and diffusion of alternative, less costly means of achieving environmental protection (Jaffe and Stavins 1995).

Strict regulatory enforcement also can lead firms simply to "go by the book" instead of searching for ongoing improvements (Bardach and Kagan 1982). Command-and-control regulation provides few direct incentives for firms to comply beyond minimal regulatory standards (Zondorak 1991). It also tends to be relatively static, whereas environmental problems and scientific understanding of these problems continuously change (Orts 1995). Promulgated through processes of debate, public comment, and negotiation, conventional forms of legislation and regulation are not easily changed once they are established. Inevitably, regulations represent only a snapshot of public concern and scientific knowledge available at a given moment. As a result, a system of public regulation may not fully encourage consistent progress toward greater environmental awareness and ever-decreasing environmental impacts.

Alternative designs for public regulatory schemes—most prominently, market-based policy instruments—offer the potential for overcoming many of the limitations of traditional forms of regulation. Market-based policies, such as emissions trading, have been implemented in several areas of environmental policy, including the phaseout of lead in gasoline and the reduction of sulfur dioxide pollution (Hahn and Hester 1989; Stavins 1989). After target emissions levels for a specific area and pollutant have been set, these policies allow firms to trade emissions permits, or average fuel content or emissions, across facilities. Some empirical studies have found that market-based approaches maintain a fixed level of environmental quality at substantially lower costs than traditional regulation (Tietenberg 1990). Although market-based instruments are extremely promising for targeted environmental problems, they have faced their own political obstacles (Keohane and others 1998) and may prove, for political and administrative reasons if no others, difficult to rely on for an extensive range of environmental problems. In addition, because emissions trading ultimately is based on public standards, it, like traditional regulation, can address only a fixed set of defined problems and may not be capable of responding any more quickly to new concerns and scientific knowledge.

Regulating from the Inside: EMSs and Their Potential for Society

EMSs may offer an additional approach to overcoming some of the limitations of the existing regulatory regime (Reiley 1997). When organizations regulate themselves from the inside, the locus of decisionmaking rests at the institutional level with the most information about the processes, technologies, and resources needed to make improvements. Internal controls also provide firms with flexibility in deciding how to minimize or reduce their environmental impacts. For these reasons, we might predict that the strategies that firms select will be more cost-effective than strategies required by uniform rules. In addition, many EMS standards encourage firms to search for strategies that both prevent pollution and save money. They also encourage firms to work with local communities in selecting environmental goals and strategies, thus creating the possibility that firms will address a broader array of their environmental impacts and make improvements that go beyond compliance with existing regulations. Firms that implement EMSs can succeed in finding ways to make environmental improvements that surpass what they otherwise would have achieved simply by following government-imposed rules.

It is easy to paint an optimistic portrait of the effects of EMSs, just as it is possible to make a pitch for almost any new management trend or policy idea. The larger and more relevant challenge is to develop an explanation for why the effects of EMSs should be positive and then to gather empirical evidence to test whether these effects occur and whether the expected explanation can be confirmed. What is it about EMSs that would enable them to improve existing environmental performance? To assess better whether managers tend to use EMSs successfully to integrate environmental concerns into business decisions, and under what conditions, it helps to understand why EMSs might contribute to better, more cost-effective environmental performance.

Systematic Management Can Yield Positive Environmental Results

One reason to expect that EMSs will yield positive results is that systematic efforts tend to yield better results than nonsystematic or haphazard efforts. As became clear to the environmental manager at LP, a systematic approach to environmental management can be an important tool for corporate managers to use to ensure that individual facilities meet their environmental responsibilities. The company found that employees at different facilities would use different, sometimes inefficient, methods of waste disposal. Since LP implemented clear standard operating procedures companywide, employees at different facilities reportedly handle waste in a consistent manner that "reduces environmental risk for local communities and increases efficiency in company operations" (Louisiana-Pacific 1999).

EMSs build on the concept of total quality management, a term coined by industry consultant W. Edwards Deming in the 1950s (Elliott 1994). Total quality management requires managers to work to continuously improve not only their products but also the processes through which products are developed, manufactured, serviced, and disposed. Total quality management means that all workers in a company are involved in pursuing the same goal: total satisfaction of customers' needs (Perigord 1990). Although total satisfaction never can be fully achieved, total quality management offers a system to move closer and closer to this goal. This system, sometimes called the "plan–do–check–act" cycle, is central to almost all EMS standards.

Some scholars and some business leaders hold the view that pollution is a kind of defect in the manufacturing process (Porter and van der Linde 1995a). Under this view, no firm intends to produce toxic benzene gas in its air emissions or heavy metals in its wastewater effluent. EMSs, based on the same approach that managers have successfully used to identify and eliminate quality defects, can help to identify and correct pollution defects. Like defects in product or service quality, pollution under this view is not an inherent by-product of economic activity but can be systematically rooted out. Therefore, organizations that set out to manage environmental matters systematically can be expected to perform better than firms that do not.

Managers Who Adopt EMSs Discover Win–Win Opportunities

EMSs may also increase value for shareholders while improving environmental performance. Some scholars argue that managers who work aggressively to prevent pollution gain an advantage over competitors (Porter and van der Linde 1995a, 1995b). Hart and Ahuja (1996) report that programs to reduce emissions led to increased profits within two years of initiation among the 127 large firms in their sample. If these studies are borne out, managers adopting ambitious waste reduction targets as part of their EMSs can expect to achieve win–win gains. As managers reduce waste, they simultaneously will improve product quality and process efficiency.

Hart emphasizes that sustainable business strategies based on pollution prevention, product stewardship, and clean technology "make good business sense" because they are sustainable over the long term (Hart 1997, 76). When "greening [is] linked to strategy or technology development," companies may realize "opportunities of potentially staggering proportions" (Hart 1997, 68). EMSs can help managers stay on track toward meeting strategic business goals.

EMSs Can Change the Culture of Firms

Some observers believe that deep and permanent improvement in environmental performance depends on change in the cultural structures of a firm (Argyris and Schon 1978; Schein 1984; Giddens 1994). Some evidence suggests that the process of developing and implementing an EMS may lead to changes in corporate culture. The link between EMS adoption and cultural change, if confirmed, would be significant—assuming, of course, that cultural change does indeed lead to important effects on environmental outcomes.

The potential of EMSs to change culture may be particularly strong when these systems are adopted as part of an industry trade association. Gunningham and Rees (1997) argue that when groups of companies join together to regulate their collective action, new norms can become institutionalized. Institutional forces within the group act to shape what members consider "appropriate" conduct through normative and mimetic means (DiMaggio and Powell 1991). Other authors argue that trade association programs such as Responsible Care may be creating and codifying new values and norms that are penetrating into the structures of participating firms, changing their preferences and thereby their routines (Nash and Ehrenfeld 1997). For example, the product stewardship code that is part of Responsible Care calls on managers to take responsibility for the environmental impacts of their products as they are transported, resold, used, and disposed of. The code states that managers must sever a business relationship with a customer that does not use the firms' products safely. This requirement extends the managers' accountability beyond the activities inside their own plant to the entire life cycle of their products. Moreover, Responsible Care's community awareness and emergency response code requires chemical manufacturing managers to include in their decisions the views of stakeholders generally neglected by an industry that traditionally has held scientific and technical expertise as the primary basis for decisionmaking.

EMSs also may change power relationships in some firms (Ehrenfeld 1998). In some Responsible Care plants, for example, responsibility for environmental concerns is often shared among managers in sales, marketing, distribution, and product development along with managers in the environmental, legal, and community relations divisions (Howard and others 2000). Because of their firms' EMSs, plant managers may have added authority to resist proposals that would be environmentally detrimental.

EMSs adopted under ISO 14001 also may increase consciousness about the environment among participating managers and instill new ways of thinking. The assumption that underlies ISO 14001 is that the environment is a management concern that must be addressed in the same systematic manner as business issues such as cost and quality. ISO 14001 requires top managers to participate in a process of continually improving a facility's EMS. It also requires that managers identify the significant environmental aspects of their activities. These requirements have the potential to spark a process of reflection about where managers see the firm moving in the environmental area. They also could lead to new consciousness about the magnitude of impacts and where they occur.

Will EMSs Really Make Much Impact?

The reasons for expecting that EMSs can bring about positive change are appealing. Systematic management leads to better environmental outcomes than haphazard management. EMSs allow managers to identify changes that improve both environmental and business performance. EMS adoption may change the culture of firms by creating a new awareness of the relationship between business activity and the environment. Some early empirical research seems to back up these arguments, but there are also reasons why private and public decisionmakers might be skeptical about the potential for EMSs, at least initially. After all, the field of public policy and management is littered with reform proposals that advocates believed would make marked improvements over the status quo, only to discover later that the perform-ance of these initiatives fails to meet expectations (for example, Coglianese 1997).

Win–Win Opportunities May Be Rare

As Palmer and others (1995) argue, unrealized opportunities for cost-saving environmental improvements may not be numerous given existing organiza-tional structures and business strategies. After all, they suggest, if the ground were littered with dollar bills, managers would have already noticed them and picked them up. At most companies, the costs of making additional environmental improvements are rapidly escalating, not diminishing (Walley and Whitehead 1994). There may be some more "low-hanging fruit" that managers can easily "pick" in the process of implementing an EMS, but such gains ultimately will be overshadowed by the total costs of firms' environ-mental programs. It may be the rare firm that reaps financial gains from its expenditures on environmental improvements. If this is correct, firms will still need to confront "real economic costs" (Walley and Whitehead 1994, 46) to make significant strides in environmental performance.

Substantial Cultural Change in Organizations Is Difficult

Even with the implementation of EMSs, firms also may confront significant obstacles in effectuating cultural change. Genuine, lasting cultural change is difficult to bring to any organization. Research in organizational behavior tends to show that changing underlying organizational values and estab-lished patterns of behavior can be a slow process (Schein 1992). Significant organizational change often requires challenging employees to abandon their old values without undermining productivity and morale. Such change may also entail challenging existing patterns of specialization and knowledge within the organization, requiring new sharing of decisionmaking authority over domains that previously had been assigned to specialists focused on the environment or on production, but not on both. Although these obstacles do not make changing organizational culture impossible, they may require an exceptional kind of organizational leadership that is not readily found nor easy to create.

Although some EMS standards contain aspirations that would seem to call for significant cultural change within organizations (for example, Responsible Care and ISO 14001), a closer reading of these standards reveals that they do not require firms to make dramatic changes or to abandon old ways of thinking about environmental responsibility. Most EMS standards accept, perhaps quite reasonably, that hazards and risks are part of doing business. Although many EMS standards pay homage to the ideals of pollution prevention, hazardous waste treatment and disposal facilities remain acceptable under the terms of most of these guidelines. Indeed, very little in a simple plan–do–check–act approach seems to call for major shifts in organizational values. Many firms may use EMSs to simply document current practices, not transform them.

Indeed, in certain instances, the formalization achieved through EMS implementation may tend to lock in existing practices. Managers may be reluctant to introduce change in facilities where formal EMS procedures have been carefully documented, every worker has been trained, and third-party registrars have certified the system. The EMS may in such cases become "an unchangeable fact" (Rikhardsson and Welford 1997, 53). Although most proponents of EMSs would argue that this possibility exists only if firms fail to implement their management systems properly, the reality is that many firms will probably implement EMSs in ways that fall short of the ideal.

Finally, even if EMSs can be expected to lead to cultural change within many firms, it is not necessarily clear how important cultural change is in producing tangible environmental improvements. Preliminary results from a thoughtfully designed empirical study of the use of EMSs in the pulp and paper industry suggest that factors external to the firm (regulation, market pressure, and community demands) may be the most important determinants of corporate environmental performance (Thornton and others 2000). If cultural change proves both difficult to produce and of comparatively little significance in improving outcomes, then the value of the EMS could be much less significant than its proponents have suggested.

Some Managers May Use EMSs to Avoid Regulatory Scrutiny

One final caution that can be raised is that some managers may adopt and use EMSs strategically to influence regulators, in an effort to preempt regulation, reduce the level of government monitoring, or prompt regulators to raise competitors' costs (Lyon and Maxwell 1999). We know that for public relations purposes, firms also may adopt EMSs opportunistically in their relationships with regulators, seeking to lead government to focus on other firms (Howard and others 2000). Like other voluntary measures (Welch and others 2000), EMSs ultimately could be used to reduce political pressure for more stringent regulation. If external regulation is a key determinant of socially beneficial environmental performance, then such an outcome—if it came to pass—would constitute a significant drawback to EMSs.

EMSs and the New Policy Agenda

Because the overall effects of EMSs are by no means guaranteed to be positive, empirical research is needed to assess the competing claims about the impact of EMSs. Are EMSs the key to fundamental improvement of industry's impact on the environment? Or are they instead a strategic cover for firms seeking to escape additional regulatory scrutiny? The true impact of EMSs almost certainly lies somewhere between the two extremes, but we will need systematic evidence to determine the proper place of EMSs within environmental policy overall.

Policymakers seldom wait for new ideas to be fully studied before taking action (Kingdon 1984). The environmental policy community's reaction to EMSs has been no exception to this trend. EMSs have found their way onto the new policy agenda in Washington, DC, and elsewhere. Legislatures and regulatory agencies across the country are currently considering and implementing programs designed to encourage firms to adopt EMSs. There has been an explosion of programs in the United States that offer financial and regulatory incentives to firms that implement EMSs (Crow 2000).

For example, EPA recently launched the National Environmental Performance Track, which seeks to motivate firms to implement EMSs and achieve higher levels of environmental performance in exchange for various promised benefits. Performance Track draws on several earlier EPA programs—Project XL (for excellence and leadership), the Environmental Leadership Program, and StarTrack—and parallels so-called green tier programs now found in about a dozen states. Each program is grounded in a desire to encourage firms to achieve high levels of environmental performance, and one common criterion used to determine whether a firm has reached that level is the establishment of an EMS. Although the features of incentive programs vary across jurisdictions, firms that implement EMSs are commonly offered greater choice in how standards are met, reduction in government oversight, penalty mitigation, expedited permitting, reduced inspection frequency, more cooperative relationships with regulators, and public recognition.

All of these policy initiatives are premised on the assumption that EMSs make a difference in environmental performance. Yet this question merits research and evidence rather than untested optimism. To their credit, the states and federal government are engaged in numerous pilot projects through which needed evidence is beginning to emerge. However, it is important to recognize that to determine whether EMSs have positive effects, researchers must untangle the effects of these systems from other factors that also may contribute to environmental improvement. Positive environmental outcomes achieved by firms with EMSs may certainly stem from the EMS itself, but they also may stem from other factors or features of the firms.

Firms make environmental improvements—even those that are costly or that exceed legal requirements—for several reasons (Reinhardt 2000). Some organizations make strides in environmental performance to stay in compliance and keep ahead of increasingly stringent regulatory standards. Others seek cost savings from more efficient use of resources. Still others seek to garner a reputation as an environmental leader. Many firms are already leaders in other areas of innovation, such as technology, which in itself may improve environmental performance (Florida 1996). Still others may seek environmental excellence in the hope of gaining some consumer advantage. Managers may simply believe that improved environmental performance is good for society and that serving the larger social welfare is part of their responsibility. When it comes to crafting public policy with EMSs in mind, it is important to know which of these several explanatory factors contribute most to improved environmental performance and under what circumstances they do. If factors other than EMSs play the key role in improving environmental performance, then public policy should support and promote those other factors and not focus on EMSs.

An EMS is a tool, something that could improve a firm's environmental performance. As with any tool, an EMS may not be sufficient to create a significant change. After all, a painter needs a brush to apply paint to canvas, and a writer needs a pen, pencil, typewriter, or computer to record words. But we do not for a moment think that the greatness of a painting or book derives from the quality of the paintbrush or the writing implement. What distinguish great painters and novelists from amateurs are not their tools so much as their skill, dedication, and perseverance. Similarly, firms that achieve great strides in pollution prevention and other improvements in environmental performance may owe little or none of that success to the mere use of an EMS. A well-implemented EMS may be one useful tool by which managers achieve, monitor, and document these improvements.[4] Improvements may depend much more on how effectively and ambitiously an EMS is implemented, how well the organization is managed overall, and how committed the managers are to seeing that the firm achieve real and continuous environmental improvement.

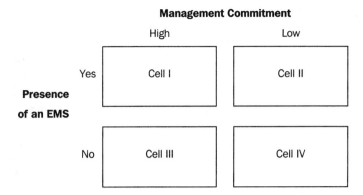

Figure 1-1. Differentiating Firms Based on Factors that Affect Environmental Performance

To understand better the impact that EMSs have on firms' environmental performance, we should distinguish any improvements caused by the EMS itself from improvements caused by factors other than the EMS. Figure 1-1 illustrates the relationship between an EMS and other factors. The term *management commitment* refers to the overall priority that a firm's top management gives to environmental improvement, whether due to market considerations, enforcement threats, or other reasons. Management commitment can be considered a proxy for various factors that contribute to environmental improvements other than the management system itself. Firms that fall into Cell I presumably will make environmental improvements, whereas those in Cell IV presumably will not. Hence, if researchers collected data only from firms in these two categories, then it would not be possible to sort out whether the EMS (or lack thereof) or the firm's commitment (or lack thereof) explained the outcome.[5] However, if researchers collected data on firms in Cells II and III, then it might then be possible to identify whether the EMS or management commitment best explained environmental improvements. If firms with high commitment but no EMS (Cell III) still made environmental progress, then we could infer that the presence of an EMS was not the important explanatory variable. Of course, a problem with Cell III is that it is hard to imagine any firm that has a high environmental commitment but does not also have some kind of management system that addresses environmental impacts, even though the system might not be called an EMS or might not necessarily meet existing EMS standards. On the other hand, if firms with low environmental commitment adopt EMSs (Cell II) and tend to make environmental improvements, then we can infer that EMSs make a difference. However, if, on average, Cell II firms did not improve as much as firms in Cell I, then we might infer that commitment still matters more.

For these reasons, evidence of progress made by firms that have volunteered to implement EMSs does not mean that EMSs are strong predictors of progress among a wider range of firms. Firms that volunteer to adopt an EMS do so because of some preexisting commitment to improving their environmental performance and therefore are less likely to implement these systems merely in token ways. For example, LP adopted its EMS to ensure the company's compliance with environmental regulations and prevent future issues such as the criminal sanctions that arose from problems at one of its facilities.

Policymakers should distinguish between the effects caused by an EMS and the effects caused by other factors, such as management commitment, for at least three reasons. First, firms that adopt and use EMSs but have a low commitment to making environmental improvements will implement their EMSs only in token or ritualistic ways, doing the bare minimum needed to be considered a firm that uses an EMS without making any significant environmental improvements (Gunningham and Grabosky 1998; Nash and Ehrenfeld 1997).[6] If a significant percentage of firms approach their EMSs in this way, then it will be less important to design public policy toward increasing the number of firms that adopt EMSs per se. It also would be wrong to rely on the presence of an EMS as an indicator of a firm's overall corporate responsibility in making decisions about how to allocate government monitoring and enforcement resources. In other words, the mere adoption of an EMS may be a weak predictor of a firm's environmental progress.

Second, the likelihood that firms will implement EMSs with different degrees of impact means that policymakers might do well to consider making any new incentives contingent on a showing that a firm exhibits the key characteristics that affect performance, such as management commitment, rather than on the mere adoption of an EMS. Because it may prove costly or difficult for government officials to obtain valid measures of factors such as management commitment, government may need instead to demand a demonstrated showing of environmental improvement. Firms could be expected to use approved tools for measuring, monitoring, and verifying actual environmental performance—and perhaps even to meet substantive performance targets—for government to justify exempting firms from existing regulatory requirements. An intermediate measure of management commitment (being used in the new EPA National Environmental Performance Track) is the commitment to quantifiable goals for environmental performance improvement beyond regulatory requirements, over specified time periods, manifested and implemented through an EMS. Of course, the more that government requires by way of a showing of eligibility for special incentives, the less attractive it will become for firms to participate in any such incentive programs. The voluntary environmental programs that have attracted the most firms appear to have been those that have placed the least demands on firms to document their performance or demonstrate their environmental commitment (Davies and Mazurek 1996). Not surprisingly, participation has been low in many of the new green tier or performance track programs adopted at the state and federal level in recent years.

Finally, if factors independent of management systems themselves are important to improving firms' environmental performance, then policymakers will need to ensure that any policies adopted to encourage the use of EMSs do not detract from these independent factors. Some policies that might dramatically increase the use of certain EMSs could unintentionally weaken other key factors, such as the motivation of management to seek environmental improvement. Policies that mandated the use of EMSs, for example, would likely lead to a formal increase in the number of firms with such systems, but the process of compulsion might nevertheless fail to promote (and could even hurt) earnest efforts by firms to look for ways to go beyond compliance with existing regulations.[7]

Overview of This Book

As we have suggested, there is reason to expect that EMSs will have important impacts on overall environmental quality but also reason to be skeptical of at least the strongest claims of EMS advocates. To untangle the effects of EMSs, systematic empirical analysis will be needed. The purpose of this book is to begin to bring such analysis to bear in an effort to inform public policy decisionmaking about EMSs. The chapters reflect the findings of a growing number of researchers who are investigating the impacts of EMSs and reporting the results of several of the most extensive research efforts to date to evaluate whether, and under what conditions, EMSs yield environmental and economic gains.

This book is organized into two parts. In Part 1, authors consider motivations (why firms choose to implement EMSs) and impacts (the results that these systems achieve). Overall, the research reported here reveals much variation in the kinds of firms that have adopted EMSs, the kinds of EMSs that these firms have adopted, the purposes for which firms use EMSs, and the effects of EMSs. As yet, no unqualified basis exists for concluding that EMSs measure up to the high expectations of their proponents; however, there are definite indications that many firms use these systems successfully to find new ideas for making socially desirable environmental improvements. Our goal is to chart a course toward a better understanding of the impacts of

EMSs to inform policy decisionmaking about whether and how government ought to respond to the growing use of EMSs by the private sector.

Part 2 focuses on the public policy implications of EMSs. Should public policy encourage—or even require—the use of EMSs? Will EMSs hold the key to a dramatic redesign of environmental policy based on performance management and principles of sustainability? As we have seen, state and federal government officials are currently considering a range of policy tools that could be used to encourage firms to develop internal management systems, such as technical assistance, public recognition, and regulatory or enforcement waivers. They also are considering proposals under which government would rely on the presence of and information generated by a private firm's EMS as alternatives to conventional regulation. Underlying this deliberation is the still deeper question of whether and why government should become involved at all in encouraging or relying on EMSs, something on which the contributors to this book certainly do not all agree.

So far, the innovative programs that EPA and the states have established have attracted only a few participants and have achieved rather modest overall gains at best. Yet these programs may turn out to be "small steps" in a larger process of "adaptive management" that leads to larger change in environmental policy (Susskind and Secunda 1998, 84). The trial-and-error efforts of state and federal governments may not ultimately yield findings that would justify a complete rewriting of the environmental regulatory system, but such experiences, especially when carefully studied, nevertheless will inform future drafts (Fiorino 1996).

The potential for EMSs to improve environmental performance and fill the gaps in our existing system of environmental regulation is promising, to be sure, but also uncertain. EMSs could perhaps herald a new wave in environmental management and policy, but such a conclusion will require additional research beyond that provided in this book. The analysis provided in this book suggests that, at best, such systems can play an important, but complementary, role in public environmental policy. In this book, we seek to lay the groundwork for understanding the effects of EMSs and how public policy can maximize their beneficial effects to give decisionmakers the opportunity to respond in realistic and effective ways to the emergence of widespread, systematic private environment management.

Acknowledgements

We thank Richard N.L. Andrews, Roy Ash, Shelley H. Metzenbaum, William R. Moomaw, Granger Morgan, and Jerry Speir for comments on an earlier draft of this chapter.

Notes

[1]Before June 2000, the American Chemistry Council was known as the Chemical Manufacturers Association.

[2]The credibility of the ISO 14001 registration process is nevertheless the subject of a forthcoming study by the National Academy of Public Administration. Questions have been raised about the consistency of interpretations of the standard by registrars and about the accountability of registrars to the public (Switzer and Ehrenfeld 1999).

[3]Particulates also have dropped, but because EPA focused on finer particles in 1987, measures of improvement for the period since 1970 are not available (EPA 2000a).

[4]Both the presence of the management system and factors other than the management system may lead firms to make environmental improvements. In short, there may be an interaction effect between environmental management systems (EMSs) and a certain kind of managerial commitment. Furthermore, the initial cause leading to an organization's decision to undertake an EMS or seek certification of its EMS need not be the same as the cause of continuing environmental improvement within the firm. Even though something other than the EMS itself causes an organization to adopt an EMS, it is plausible that the EMS and its accompanying institutional norms and structures help sustain management's initial commitment. The initial commitment by management to make environmental improvements by adopting an EMS, for example, may become entrenched by the EMS itself, helping to carry forward the initial commitment even at later moments, when management otherwise might not have pursued environmental improvement.

[5]Firms that adopt EMSs probably also possess a high level of commitment to making environmental improvements (Cell I). This duality presents a major challenge for research on the impact of EMSs, because currently most firms have volunteered to implement EMSs and none are assigned randomly to implement them. Firms that volunteer are more likely to be firms that would have improved their environmental performance anyway, which will inherently limit what can be learned from studies of firms that volunteer to implement EMSs (Rogers 1995).

[6]Interestingly, the definition of *continual improvement* under ISO 14001 does not necessarily dictate an improvement in the *environmental impact* of the organization (that is, the "change to the environment ... wholly or partially resulting from an organization's activities, products or services") but rather in *environmental performance,* which is defined as "measurable results of the environmental management system, related to an organization's control of its environmental aspects, based on its environmental policy, objectives and targets" (ISO 14001, Part 3, Environmental Management Systems, 1996). The kind of "measurable results" that are to be improved could certainly be results in terms of environmental impacts, but presumably they also could be results in terms of cost savings or other results for the organization itself instead of for society or the environment.

[7]As Kagan (1998) has argued, whereas enforceable mandates or what he calls "deterrence models" may work well "at inducing literal compliance, or *accountability* according to law, they may tend to undercut the continuing exercise of *responsibility*

and improvements in performance that the best self-regulatory systems generate" (emphasis in original).

References

Ackerman, B., and R. Stewart. 1985. Reforming Environmental Law. *Stanford Law Review* 37: 1333–1365.

AF&PA (American Forest and Paper Association). 2000. Sustainable Forestry Initiative. http://afandpa.org (accessed June 30, 2000).

American Chemistry Council. 2000. Responsible Care Guiding Principles. http://AmericanChemistry.com (accessed June 30, 2000).

Argyris, C., and Donald A. Schon. 1978. *Organizational Learning: A Theory of Action Perspective.* Reading, MA: Addison-Wesley.

The Aspen Institute. 1996. *The Alternative Path: A Cleaner, Cheaper Way to Protect and Enhance the Environment.* Washington, DC: The Aspen Institute.

Ayres, Ian, and John Braithwaite. 1992. *Responsive Regulation: Transcending the Deregulation Debate.* New York: Oxford University Press.

Balta, Wayne, and Gayle Woodside. 1999. Try It—You'll Like It. *The Environmental Forum* March/April: 36.

Bardach, Eugene, and Robert A. Kagan 1982. *Going by the Book: The Problem of Regulatory Unreasonableness.* Philadelphia, PA: Temple University Press.

Bok, Derek. 1996. *The State of the Nation: Government and the Quest for a Better Society.* Cambridge, MA: Harvard University Press.

Breyer, Stephen G. 1993. *Breaking the Vicious Circle: Toward Effective Risk Regulation.* Cambridge, MA: Harvard University Press.

Carter, Adrian. 1999. Integrating Quality, Environment, Health and Safety Systems with Customers and Contractors. *Greener Management International* 28: 59–68.

Chertow, Marian, and Daniel C. Esty (eds.). 1997. *Thinking Ecologically: The Next Generation of Environmental Policy.* New Haven, CT: Yale University Press.

Coglianese, Cary. 1997. Assessing Consensus: The Promise and Performance of Negotiated Rulemaking. *Duke Law Journal* 46: 1255.

Crow, Michael. 2000. Beyond Experiments. *Environmental Forum* May/June: 19–29.

Davies, J. Clarence, and Jan Mazurek. 1998. *Pollution Control in the United States: Evaluating the System.* Washington, DC: Resources for the Future.

Davies, Terry, and Jan Mazurek. 1996. *Industry Incentives for Environmental Improvement: Evaluation of U.S. Federal Initiatives.* Report to the Global Environmental Management Initiative. Washington, DC: Resources for the Future.

DiMaggio, P., and W. Powell (eds.). 1991. *The New Institutionalism in Organizational Analysis.* Chicago, IL: University of Chicago Press.

Ehrenfeld, John R. 1998. Cultural Structure and the Challenge of Sustainability. In *Better Environmental Decisions: Strategies for Governments, Businesses, and Communities,* edited by Ken Sexton, Alfred Marcus, K. William Easter, and Timothy D. Burkhardt. Washington, DC: Island Press.

Elliott, E. Donald. 1994. Quality Environmental TQM: Anatomy of a Pollution Control Program that Works! *Michigan Law Review* 92: 1840–1854.

Elliott, E. Donald, and Gail Charnley. 1998. Toward Bigger Bubbles. *Forum for Applied Research and Public Policy* 13: 48–54.

Fiorino, Daniel J. 1996. Toward a New System of Environmental Regulation: The Case for an Industry Sector Approach. *Environmental Law* 26: 457–488.

———. 1999. Rethinking Environmental Regulation: Perspectives on Law and Governance. *Harvard Environmental Law Review* 23: 441.

Fischer, Kurt, and Johan Schot (eds.). 1993. *Environmental Strategies for Industry*. Washington, DC: Island Press.

Florida, Richard. 1996. Lean and Green: The Move to Environmentally Conscious Manufacturing. *California Management Review* 39(1): 80–105.

Furrer, Bettina, and Heinrich Hugenschmidt. 1999. Financial Services and ISO 14001: The Challenge of Determining Indirect Environmental Aspects in a Global Certification. *Greener Management International* 28: 32–41.

Giddens, Anthony. 1994. *The Constitution of Society*. Berkeley, CA: University of California Press.

Graham, John, and Jonathan Wiener (eds.). 1995. *Risk versus Risk: Tradeoffs in Human Health and Environmental Protection*. Cambridge, MA: Harvard University Press.

Gunningham, Neil, and Peter Grabosky. 1998. *Smart Regulation: Designing Environmental Policy*. Oxford: Oxford University Press.

Gunningham, Neil, and Joseph Rees. 1997. Industry Self-Regulation: An Institutional Perspective. *Law and Policy* 19(4): 363–414.

Hahn, Robert W. 1996. *Risks, Costs, and Lives Saved: Getting Better Results from Regulation*. New York: Oxford University Press.

Hahn, Robert W., and Gordon L. Hester. 1989. Marketable Permits: Lessons for Theory and Practice. *Ecology Law Quarterly* 16: 361–406.

Hart, Stuart L. 1997. Strategies for a Sustainable World. *Harvard Business Review* 75(1): 66–76.

Hart, Stuart L., and Gautam Ahuja. 1996. Does It Pay to be Green? *Business Strategy and the Environment* 5: 30–37.

Howard, Jennifer, Jennifer Nash, and John Ehrenfeld. 2000. Standard or Smokescreen? Implementation of a Voluntary Environmental Code. *California Management Review* 42(2): 63–82.

Howes, Rupert, Jim Skea, and Bob Whelan. 1997. *Clean and Competitive? Motivating Environmental Performance in Industry*. London, U.K.: Earthscan Publications Ltd.

Jaffe, Adam, and Robert Stavins. 1995. Dynamic Incentives of Environmental Regulations: The Effects of Alternative Policy Instruments on Technology Diffusion. *Journal of Environmental Economics and Management* 29(3): S43–S63.

Johnston, Craig N. 1995. An Essay on Environmental Audit Privileges: The Right Problem, the Wrong Solution. *Environmental Law* 25: 335–347.

Kagan, Robert A. 1998. American Adversarial Legalism and Intra-Corporate Regulatory Systems. Paper presented at the annual meeting of the Law and Society Association, Aspen, Colorado.

Keohane, Nathaniel, Richard Revesz, and Robert Stavins. 1998. The Choice of Regulatory Instruments in Environmental Policy. *Harvard Environmental Law Review* 22: 313–367.

Kingdon, John W. 1984. *Agendas, Alternatives, and Public Policies*. New York: Harper Collins.

Kleindorfer, Paul R. 1999. Understanding Individuals' Environmental Decisions: A Decision Science Approach. In *Better Environmental Decisions: Strategies for Governments, Businesses, and Communities*, edited by Ken Sexton, Alfred Marcus, K. William Easter, and Timothy D. Burkhardt. Washington, DC: Island Press.

Louisiana-Pacific Corporation. 1999. *Louisiana-Pacific Environmental Management System*. Portland, Oregon: Louisiana-Pacific Corporation.

Lyon, Thomas P., and John W. Maxwell. 1999. "Voluntary" Approaches to Environmental Regulation: A Survey. In *Environmental Economics: Past, Present and Future*, edited by Maurizio Franzini and Antonio Nicita. Aldershot, Hampshire, U.K.: Ashgate Publishing Ltd.

Meidinger, Errol E. 1999–2000. "Private" Environmental Regulation, Human Rights, and Community. *Buffalo Law Review* 7: 125–237.

Nash, Jennifer, and John R. Ehrenfeld. 1997. Codes of Environmental Management Practice: Assessing Their Potential as Tools for Change. *Annual Review of Energy and Environment* 22: 487–535.

Orts, Eric. 1995. Reflexive Environmental Law. *Northwestern University Law Review* 89: 1227–1290.

Palmer, Karen, Wallace E. Oates, and Paul R. Portney. 1995. Tightening Environmental Standards: The Benefit-Cost or the No-Cost Paradigm? *The Journal of Economic Perspectives* 9(4): 119–132.

Perigord, Michel. 1990. *Achieving Total Quality Management: A Program for Action*. Cambridge, MA: Productivity Press.

Pildes, Richard H., and Cass R. Sunstein. 1995. Reinventing the Regulatory State. *University of Chicago Law Review* 62: 1.

Porter, Michael E., and Claas van der Linde. 1995a. Green and Competitive: Ending the Stalemate. *Harvard Business Review* 73: 120–134.

———. 1995b. Toward a New Conception of the Environment-Competitiveness Relationship. *Journal of Economic Perspectives* 9(4): 97–118.

Portney, Paul R. 2000. EPA and the Evolution of Federal Regulation. In *Public Policies for Environmental Protection* (Second Edition), edited by Paul R. Portney and Robert N. Stavins. Washington, DC: Resources for the Future.

Portney, Paul R., and Robert N. Stavins (eds.). 2000. *Public Policies for Environmental Protection* (Second Edition). Washington, DC: Resources for the Future.

Rees, Joseph. 1997. The Development of Industry Self-Regulation in the Chemical Industry. *Law and Policy* 19(4): 447–528.

Reiley, Robert Anthony. 1997. The New Paradigm: ISO 14000 and its Place in Regulatory Reform. *Journal of Corporation Law* 22: 535.

Reinhardt, Forest L. 2000. *Down to Earth: Applying Business Principles to Environmental Management*. Boston, MA: Harvard Business School Press.

Rikhardsson, Pall, and Richard Welford. 1997. Clouding the Crisis: the Construction of Corporate Environmental Management. In *Hijacking Environmentalism: Corporate Responses to Sustainable Development*, edited by Richard Welford. London, U.K.: Earthscan Publications Ltd.

Rogers, Everett. 1995. *Diffusion of Innovations* (Fourth Edition). New York: Free Press.

Schein, Edgar. 1984. Coming to a New Awareness of Organizational Culture. *Sloan Management Review* 25(2): 3–16.

Schein, Edgar H. 1992. *Organizational Culture and Leadership* (Second Edition). San Francisco, CA: Jossey-Bass.

Stavins, Robert. 1989. Harnessing Market Forces to Protect the Environment. *Environment* 31: 28–35.

Susskind, Lawrence, and Joshua Secunda. 1998. The Risks and the Advantages of Agency Discretion: Evidence from EPA's Project XL. *UCLA Journal of Environmental Law and Policy*. 17: 67–116.

Switzer, Jason, and John R. Ehrenfeld. 1999. Independent Environmental Auditors: What Does ISO 14001 Registration Really Mean? *Environmental Quality Management* Autumn: 17–33.

Thornton, Dorothy, Robert A. Kagan, and Neil Gunningham. 2000. In Search of Environmental Leadership: Law, Community, Market, and Management in the Pulp and Paper Industry. Paper presented at the Annual Meeting of the Law and Society Association, Miami, FL.

Tietenberg, Tom H. 1990. Economic Instruments for Environmental Regulation. *Oxford Review of Economic Policy* 6: 17–33.

U.S. EPA (Environmental Protection Agency). 2000a. *Latest Findings on National Air Quality: 1999 Status and Trends*. EPA-454/F-00-002. Office of Air Quality Planning and Standards. Research Triangle Park, NC: U.S. EPA.

———. 2000b. *1998 Toxic Release Inventory: Public Data Release*. September. Washington, DC: U.S. EPA.

Walley, Noah, and Bradley Whitehead. 1994. It's Not Easy Being Green. *Harvard Business Review* 72(3): 46–53.

Welch, Eric W., Allan Mazur, and Stuart Bretchneider. 2000. Voluntary Behavior by Electric Utilities: Levels of Adoption and Contribution of the Climate Challenge Program to the Reduction of Carbon Dioxide. *Journal of Public Policy Analysis and Management* 19: 407–425.

Zondorak, V.A. 1991. A New Face in Corporate Environmental Responsibility: The Valdez Principles. *BC Environmental Affairs Law Review* 18: 457–499.

Part 1

Motivations and Impacts

Why are some managers taking on the costs associated with implementing an environmental management system (EMS)? Are they motivated by external factors, such as market demand or pressure from regulators, or are they driven by their organization's internal strategy and culture? And after they adopt an EMS, does the new system bring about changes in the organization's environmental performance? Does the firm change the level of resources it devotes to environmental protection, the integration of environmental concerns into business decisions, or the ways it seeks out and uses the perspectives of external stakeholders? These questions frame the chapters presented in Part 1 of this book.

In Chapter 2, Richard N.L. Andrews and a team of researchers from the University of North Carolina at Chapel Hill and the Environmental Law Institute begin with an overview of the evolution of environmental policy in the United States and the emergence of environmental practices that go beyond regulatory requirements. They discuss theories of corporate behavior and the natural environment, explaining that until recently, environmental protection was viewed as a cost that when transferred back to the firm inevitably worsened the firm's manufacturing performance. Now some theorists are making nearly the opposite argument, namely, that focused attention on environmental management can produce competitive advantage. Andrews and others argue that the move toward greater integration of environmental concerns into the decisionmaking of firms must be evaluated not only in the context of hopeful "green" idealism but also through careful attention to the economic forces that drive and constrain business outcomes. Public policy that relies even partially on private voluntary initiatives will succeed only if the economic benefits are powerful enough to motivate and sustain the commitments of firms.

To understand the role of EMSs in environmental and economic performance, the authors created a National Database on Environmental Management Systems (NDEMS). As of June 2000, more than 50 facilities had begun to provide detailed information on many aspects of their EMSs. Interestingly, not all facilities report net economic benefits from EMS adoption, but some do. Most managers believe the effort was nevertheless worthwhile, claiming that they would do it again even though the EMS did not strictly pay for itself. It will be important to track EMSs at these facilities to see whether managers remain committed to systems that do not produce direct economic payoffs. Indeed, one of the primary objectives of the NDEMS is to understand the sustainability of EMS commitments over time and across changes in organizational ownership, structure, and management.

Andrews and others observe that managers in most of the facilities that participated in their study use EMSs to investigate all potential sources of environmental impact. They have given priority to regulatory compliance in their EMSs, raising the question of whether, over time, firms will expand their goals more broadly. Andrews and others (along with Florida and Davison in Chapter 4) find that regulatory compliance often is the primary motivator for EMS adoption among managers they have surveyed, suggesting that compliance will continue to be an EMS focus. The question remains whether in the future, EMSs will evolve from focusing on compliance-oriented goals to addressing a broader array of impacts, including unregulated emissions, resource consumption, and the impacts of products after they leave the manufacturing site.

In Chapter 3, Jennifer Nash and John R. Ehrenfeld suggest that substantial information is already available about how managers use EMSs. Individual firms have developed and used corporate governance structures with many of the features of EMSs since the mid-1980s, and numerous trade associations have developed environmental codes of management practices that serve as EMS standards. ISO 14001 shares many features in common with these approaches. Nash and Ehrenfeld argue that managers implement EMSs within the context of their firm's individual culture, strategic goals, and competitive position. Managers who see environmental practices as marginal to their strategic and competitive objectives appear to treat EMSs as tools primarily for external image manipulation. Firms with strong environmental commitments use EMSs as tools to become even stronger. In other words, EMSs serve mainly as reinforcing mechanisms. Two organizations can implement the same type of EMS but end up with very different results. To understand the likely outcome of EMS adoption, one must understand why managers were motivated to implement the EMS in the first place and how the EMS fits within the overall culture of the organization.

In Chapter 4, Richard Florida and Derek Davison ask whether "advanced plants"— plants that implement both EMSs and pollution prevention strategies—take what they have learned inside their factories and apply it to reduce their environmental impacts on the surrounding communities. Do plants that innovate in terms of their internal management also innovate when it comes to working with external constituencies? The authors surveyed nearly 500 "advanced" and "nonadvanced" facilities, concluding that advanced plants appear to pose less environmental risk and confer more significant benefits on their communities than plants without such practices. They argue that organizational resources tend to operate as a system, creating the capacity to respond to both internal opportunities and external events.

In Chapter 5, Theodore Panayotou considers the role of EMSs in the global economy. Whereas other authors in Part I address EMSs within the context of the U.S. regulatory system, Panayotou considers their value in other settings, particularly in countries where environmental regulations are weak and rarely enforced. He argues that EMSs have the potential to deliver greater value than the status quo in the developing world. An EMS provides a framework for "voluntary but systematic" procedures, without commanding specific standards. When appropriately designed and implemented, EMSs promise to advance both economic and environmental performance by taking due account of individual firm, industry, and country conditions and optimizing between standardization and flexibility. Use of EMSs in developing countries may eventually lead to a convergence of environmental practices, Panayotou claims, through the requirement for continual improvement.

Two key lessons emerge from these chapters. First, firms with strong EMSs tend to excel in multiple areas. Andrews and others (Chapter 2) focus on facilities that are cooperating with both researchers and voluntary pilot programs of their state environmental agencies. The facilities studied by Florida and Davison (Chapter 4) were chosen on the basis of their past reputation for environmental excellence. Firms that adopt ambitious EMSs tend also to prevent pollution and reach out to their communities. Nash and Ehrenfeld (Chapter 3) remind us that managers of these facilities have often already decided to pursue environmental excellence and seek to use EMSs as a tool to achieve this goal.

A second lesson is that a facility's EMS should not be viewed in isolation but as part of a larger set of management decisions. Andrews and others (Chapter 2) emphasize that EMSs must help managers achieve their business objectives if they are to have a lasting impact. Nash and Ehrenfeld

(Chapter 3) emphasize that an EMS grows out of a firm's culture, whereas Florida and Davison (Chapter 4) stress the system of organizational capabilities of which an EMS is a part. Panayotou (Chapter 5) focuses on the regulatory context of the EMS. Whether to adopt an EMS is one of many decisions managers make about how best to achieve or reinforce their business objectives, culturally derived expectations, organizational capabilities, and regulatory strategy.

Together, these chapters advance our understanding of how EMSs help organizations achieve their established objectives and strategies and lay the groundwork necessary for setting public policy with respect to EMSs. The findings in these chapters help us to understand the degree to which an EMS merely reflects an organization's established commitments or leads to a fundamental rethinking of day-to-day routines, core competencies, and external relationships.

2

Environmental Management Systems: History, Theory, and Implementation Research

Richard N.L. Andrews, Nicole Darnall, Deborah Rigling Gallagher, Suellen Terrill Keiner, Eric Feldman, Matthew L. Mitchell, Deborah Amaral, and Jessica D. Jacoby

The widespread adoption of formal environmental management systems (EMSs) by businesses and other organizations has been promoted as an innovation that has the potential to alter profoundly their environmental and economic performance and their resulting relationships with longstanding environmental regulatory policies and agencies. Advocates argue that when implemented, EMSs have the potential not only to improve compliance with environmental regulations but also to refocus the organization's attention beyond compliance toward a dynamic, continual process of improvement in environmental and economic performance. In the process, the organization likely will discover new opportunities to prevent rather than simply control pollution and opportunities to reduce wasteful uses of resources, thus saving money while improving the environment. It also may discover opportunities to manage the organization as a whole more effectively.

Many businesses have developed environmental management procedures for years, but until recently, there was no trend toward formalizing or standardizing them more generally. Often within corporations, EMSs remained largely the responsibility of a single office that oversaw regulatory compliance and risk minimization, such as the office of a vice president for environment, health and safety, rather than an organization-wide mission for which all managers would be held accountable. In late 1996, however, the International Organization for Standardization published the final version of an international voluntary EMS standard, ISO 14001. Other documents in the ISO 14000 series provide more detailed guidance on many EMS-related topics, such as life cycle analysis, eco-labeling, environmental auditing, and environmental performance evaluation.

An EMS is a formal set of policies and procedures that define how an organization will manage its potential impacts on the natural environment and on the health and welfare of the people who depend on it. The ISO 14001 standard provides an explicit and closely documented procedural template for such a system, which can be audited and certified by an approved third-party "registrar" as conforming to the ISO 14001 standard. At a minimum, organizations that adopt the ISO 14001 standard accept a responsibility to

- adopt a written environmental policy;
- identify all environmental aspects and significant impacts of their activities, products, and services;
- set objectives and targets for continuous improvement in environmental performance;
- assign clear responsibilities for implementation, training, monitoring, and corrective actions; and
- evaluate and refine implementation over time so as to achieve continuous improvement both in the implementation of environmental objectives and targets and in the EMS itself.

Similar procedural standards, varying somewhat in the details, have been adopted in Great Britain (BS 7750) and the European Union (the Eco-Management and Auditing Scheme [EMAS]).

As of July 2000, an estimated 18,052 facilities worldwide had been certified as meeting the ISO 14001 standard, including approximately 850 in the United States. The number of U.S. facilities approximately tripled between 1998 and 2000. The increasing trend of U.S. adoption of ISO 14001 was bolstered by an onset of business-driven mandates and government programs. In September 1999, the Ford Motor Company and General Motors announced intentions to require ISO 14001 certification of all their first-tier suppliers' manufacturing sites by July 2003 (Ford) and by the end of 2002 (General Motors), and to encourage them to require such certification of second- and third-tier suppliers as well. Toyota announced a similar requirement, effective by the end of 2003. In April 2000, President Bill Clinton issued an Executive Order mandating that each federal agency implement an EMS at "all appropriate agency facilities based on facility size, complexity, and the environmental aspects of facility operations" no later than December 2005 (EO 13148, April 22, 2000). Finally, the U.S. Environmental Protection Agency (EPA) and some state governments adopted policies that encourage EMS adoption and certification, establishing "performance tracks" with incentives to benefit firms that implement EMSs, and incorporating EMS requirements into some supplemental environmental projects (SEPs) for firms out of compliance with environmental regulations (http://www.epa.gov/performancetrack). For all these reasons and others, the implementation and certification of ISO 14001–compliant EMSs will likely continue to increase in the United States and worldwide.

The advent of widely used and formally documented EMSs therefore raises important questions for both research and public policy. As a research topic, it offers a unique opportunity to observe the processes and the environmental and economic consequences of these initiatives, and to compare similarities and differences across different firms, sectors, sizes, and other characteristics. From a public policy perspective, it offers an unusual opportunity to look at environmental management decisions through the eyes of the businesses that make them, not only from the perspective of the government agencies that seek to influence them. At the same time, it also should shed new light on important environmental policy questions, such as whether the environmental management initiatives of businesses themselves can produce more effective and economical environmental performance, better regulatory compliance, more efficient monitoring and reporting procedures, or other benefits to the public as well as to the firms themselves.

The importance of these questions led to the creation of the Multi-State Working Group on Environmental Management Systems (MSWG) and EPA support for creation of a National Database on Environmental Management Systems (NDEMS), in cooperation with MSWG and with more than 50 participating facilities, to provide systematic empirical evidence of the effects of EMS implementation. The NDEMS database is being developed jointly by the University of North Carolina (UNC) at Chapel Hill and the Environmental Law Institute (ELI).[1]

In this chapter, we provide historical and theoretical contexts for the initial research being carried out on this database, some preliminary results, and discussion of further research plans, needs, and opportunities.

Historical Context

For more than three decades, scholars and policy advocates have argued over what combination of voluntary measures, economic incentives, and government regulations represents the best way to control pollutant emissions and other environmental impacts (Andrews 1999). Before 1970, the dominant approach was voluntary measures, plus regulations in some states; by 1970, this approach was widely viewed as inadequate, and a series of major new federal regulatory statutes established technology-based permit requirements and other restrictions for air and water emissions and waste treatment facilities.

Critics have since denigrated these regulations as an unduly rigid and inefficient "command-and-control" approach (for example, U.S. EPA 1990). In fact, the regulations were largely successful, though costly, in significantly reducing air pollutant emissions and wastewater discharges as well as improving municipal and hazardous waste management (Andrews 1999). However, they dominated the agendas of businesses and government, producing a preoccupation with regulatory compliance rather than with full and efficient integration of environmental considerations into the core goals and decisions of businesses. Environmental management was treated as a necessary evil rather than as a business opportunity: a regulatory burden that was assigned to pollution control engineers responsible for end-of-pipe technological equipment, rather than a new core function that should be the shared responsibility of managers throughout the organization. The question remained, therefore, whether more efficient and effective means for achieving better environmental performance could be found.

Since at least the 1970s, it also has been documented that environmental impacts were themselves signals of economic inefficiency in production, which should have been corrected in the interest of industrial as well as societal optimization. Kneese and Bower (1979) documented economically efficient opportunities for pollution prevention in a series of industries during the 1970s, and Royston (1979) began popularizing the idea that "pollution prevention *pays.*" A few leading corporations (3M, for instance) also began publicizing this idea, and other studies confirmed it (for example, Sarokin and others 1985).

As regulatory enforcement tightened in the 1980s, many more businesses began instituting environmental auditing practices, initially for compliance assurance but also for due-diligence management of potential liability (especially by banks and insurance companies, in the wake of Superfund strict liability legislation for hazardous waste dumping in 1980). These practices expanded rapidly after the Bhopal industrial disaster in 1984 (Hemphill 1995) and the required public reporting of toxic pollutant releases beginning in 1986 (the Toxics Release Inventory [TRI]), which documented for the first time the actual quantities of pollutant releases by many major businesses and thus generated new incentives for the firms themselves to identify and reduce emissions (Andrews 1999). During the 1980s, many businesses also integrated environment, health, and safety (EHS) responsibilities under a single EHS vice president, moving beyond the narrow compliance technology model of the 1970s toward a more managerial approach. In 1991, EPA Administrator William Reilly began offering modest federal rewards for voluntary industrial initiatives to prevent and reduce emissions (the 33/50 and Green Lights programs); whether due to the additional incentives or not, the pollution reduction effort was an important success and increased the legitimacy of voluntary business initiatives to reduce pollution (Davies and Mazurek 1996).

Finally, in preparation for the United Nations' 1992 "Earth Summit," the World Business Council for Sustainable Development (WBCSD) issued a visionary declaration asserting the "inextricable linkage" among economic growth, environmental protection, and the satisfaction of basic human needs and calling for "far-reaching shifts in corporate attitudes and new ways of doing business" to achieve environmental and social sustainability. Significantly, the WBCSD report posed this goal squarely as a challenge and opportunity for businesses; and at its initiative, the International Organization for Standardization set up a strategic advisory group to measure "eco-efficiency," whose efforts led to the creation of the ISO 14000 series of environmental management standards (Schmidheiny 1992).

By the 1990s, in some business and government circles in the United States and worldwide, there also was active advocacy for increased self-regulation of businesses for environmental protection and pollution reduction. Specific proposals included sectoral pollution reduction "covenants," market trading of emissions permits, third-party certification of environmental performance (for example, under the European Union's EMAS and the ISO 14001 Environmental Management System certification), and ad hoc negotiation of regulatory flexibility in exchange for superior environmental performance. Environmental sustainability had been publicly adopted—at least by some leading businesses and executives—as a fundamental business goal and opportunity, and environmental management as a core business function. However, certain questions remained. How deep was this commitment? How widely shared was it among all firms and sectors? How durable would it be in the face of conflicting economic pressures, such as the seemingly relentless demand for short-term profitability?

These questions hold central importance for the ongoing public policy debate over what reforms in environmental regulatory statutes should be considered. Should environmental regulations be implemented more flexibly in exchange for voluntary actions by industries, or would such flexibility in reality open the door to endless special pleadings and to erosion of the regulatory framework that has produced much of the environmental progress to date?

Theories of Corporate Behavior and the Natural Environment

On what grounds should one expect businesses—or nonmarket organizations such as government operations, for that matter—to achieve environmental performance superior to what was required of them by regulation?

Even if they did, why should one expect them to undertake the elaborate bureaucratic costs—paperwork, process, and external audits—associated with an ISO 14001–certified EMS, rather than simply implementing cost-effective changes ad hoc?

For at least half a century, environmental problems have been characterized as externalities and commons problems, imposing social rather than business costs and thus rationally ignored by businesses until forced back on them by government action (Kapp 1950). Traditional environmental economics argued that, although prescriptive regulations are inefficient because they dictate inflexible and suboptimal means for achieving environmental goals, government should nevertheless provide market-based incentives such as emissions taxes or marketable credits to correct market signals that undervalue environmental assets (see, for example, Jaffe and others 1995).

A parallel business literature assumed a zero-sum trade-off relationship between business costs and social costs, such that any environmental improvement by a firm was assumed merely to transfer costs previously incurred by society back onto the firm, thus worsening the firm's manufacturing performance in terms of cost, quality, speed, and flexibility (Klassen and Whybark 1999, and references therein). Standard engineering and economics textbooks postulated an exponentially rising curve for pollution control costs with each increment of improved protection. Why, then, should businesses act voluntarily to internalize these costs if it is more efficient for them to push the costs off onto society in the first place?

A revisionist empirical literature began to appear in the 1980s, documenting a surprisingly pervasive range of cases in which pollution prevention investments in fact produced economic benefits for both society and the business itself (for example, Royston 1979; Sarokin and others 1985; Cairncross 1991; Porter 1991; Schmidheiny 1992; Smart 1992; Fischer and Schot 1993; Hawken 1993; Allenby and Richards 1994; Porter and van der Linde 1995a, 1995b; Hart 1997; Hawken and others 1999). Experiences of several major corporations, especially after implementation of the TRI reporting requirements, added support to this position. Far from operating at peak efficiency as assumed by economic theory, many firms appeared to be operating mainly by habit, leaving significant opportunities for cost-effective environmental improvements "lying on the table." A growing chorus of voices thus argued for a "greening of industry" that would benefit both society and the businesses themselves. This literature raised important challenges to the assumptions of traditional economic theory, but it remained largely atheoretical itself: why were businesses failing to recognize and correct such inefficiencies if it was in fact in their own economic self-interest to do so?

At the same time, a growing critical literature argued that existing approaches to environmental regulation, particularly in the United States, were at least inefficient and in some views ineffective as well. Beginning in the late 1970s, a literature developed attacking government "overregulation" and demanding regulatory reforms (see Weidenbaum 1979). This reform agenda was sidelined for several years as the Reagan administration instead attempted to dismantle federal environmental regulation, but it then reemerged in the form of an oversimplified cliché that pejoratively contrasted command-and-control regulations with the more attractive-sounding "market-based incentives" (see, for example, U.S. EPA 1990). These arguments included claims that conventional environmental regulation was inefficient, imposing higher costs than were necessary to achieve the desired environmental performance goals; that it was ineffective, by requiring end-of-pipe control technologies that merely move pollutants around (among air, water, and land) rather than reducing total releases; that it was unworkable for "new generation" environmental problems, such as the cumulative effects of many widely dispersed nonpoint and mobile sources, even if it was effective in reducing pollution from major point sources; and that it was increasingly unenforceable in any case, because governments increasingly lacked the resources, the political will, and, in an open global economy, the effective authority to enforce conventional environmental regulations.

The response to these arguments was a variety of proposals, in Europe and the United States, to encourage increased environmental self-regulation by businesses. Some of these proposals would serve as alternative implementing mechanisms for public environmental goals, standards, and licensing requirements. Others proposed fundamentally different approaches to environmental performance, claiming to be based on the enlightened self-interest and commitment of businesses themselves. Whether these proposals would in fact produce better results than existing regulatory regimes and whether existing regulatory regimes or self-regulation proposals would produce environmentally sustainable results are important questions that have not yet been systematically answered. To address them requires theoretical and empirical investigation of environmental decisionmaking in businesses, a subject that until recently has not been widely studied.

The effectiveness of environmental self-regulation mechanisms must be evaluated in the context not only of hopeful "green" idealism, or the rhetoric and anecdotes of a few leading firms and their image makers, but also of the basic economic forces that drive and constrain business outcomes and that will relentlessly, if not immediately, sort long-term trends from fleeting experiments in business decisionmaking.[2]

A privately held company can pursue whatever objectives its owners desire, within the limits of the laws, including accepting less than maximum short-term profits to be seen as a good community citizen or to maintain the long-term legacy and reputation of the firm. Publicly owned facilities also can pursue the objectives of the state that operates them, whether those be maximizing production at the expense of the environment or giving procurement preferences to green suppliers. Publicly traded businesses, however, are subject to more rigid laws of the current global marketplace. In global markets in which capital is free to move instantaneously to the most immediately profitable investments anywhere in the world, this pressure is controlled not so much by consumers as by investors. Higher-cost firms lose customers and investors, and underperforming firms are vulnerable to involuntary restructuring to serve short-term investor interests. Fundamentally, such firms are increasingly forced to pay primary attention to short-term profitability.

These market forces may threaten the substantial EHS staff capabilities that some leading businesses have developed at the corporate level. Even as some corporate EHS executives are hopeful of gaining increased influence at headquarters, others are finding themselves marginalized or even eliminated by the decentralization of decisionmaking to individual production units whose mandate is to produce short-term returns or be closed or sold.[3] Privately held firms and publicly owned facilities may experience less pressure, but they too must face cost-minimizing competitors unless they operate as monopolies or in protected markets.

Proposals for voluntary environmental self-regulation therefore must demonstrate that there are private benefits of environmental self-regulation; that these private benefits exceed the private costs of undertaking it, in the short term required by financial markets; and that these private net benefits are sufficient to motivate private environmental performance that is equal or superior to public environmental standards (R. Andrews 1998).

Research on environmental management in business is still in its infancy. Schot and Fischer (1993, 372–373), in a research agenda on environmental strategies for industry, urge that more research be devoted to in-depth case studies to determine how learning processes occur within and among organizations that lead them from a defensive to an innovative environmental management approach, using theories developed within the framework of strategic management, organizational, and innovation studies. They also called for research on how government policies can induce this transition and on the influence of evolving new relationships between firms and the environmentally concerned public, which they identified as one of the most important driving forces for changes in the behavior of firms.

One line of such research proposed that environmental innovation is driven primarily by external forces, such as regulatory or market pressures. Porter (1991) in particular argues that in practice, government regulations may serve as a stimulus to economic growth and cleaner production if production changes in response to regulation are used as a business asset to gain market advantages over competitors. A later review of the literature, however, concludes that neither positive nor negative effects of environmental regulation on competitiveness were easily detectable (Jaffe and others

1995). Porter and van der Linde (1995a, 1995b) conclude that firms seek to maximize "resource productivity" in response to regulatory and market pressures, enabling them to simultaneously improve industrial and environmental performance (Florida and others 1999).

However, others continue to challenge this conclusion (Palmer and others 1995). An alternative line of theory proposes that the economic and environmental performance of businesses is driven by primarily internal forces, including management strategy and firm-level resources (Hart 1995; Klassen and McLaughlin 1996; Klassen and Whybark 1999). This resource-based view of the firm postulates that sustained competitive advantage is driven by the firm's use of strategic resources—its financial assets and capabilities and its less tangible knowledge-based advantages, such as socially complex organizational processes and reputational assets. The latter are rare, difficult to imitate, and have few substitutes.

In an early and insightful article on this subject, Gabel and Sinclair-Desgagné (1994) propose that poor environmental management is caused not only by market or regulatory failures, with which environmental economists and policy scholars are preoccupied, but by organizational failures on the part of businesses themselves. Framing their argument from the perspective of principal–agent theory, Gabel and Sinclair-Desgagné argue that businesses often recognize the value of environmental goals in principle but fail to operationalize them throughout the management systems that drive employees' behavior: the compensation system, quantification and monitoring of nonfinancial objectives, internal pricing, horizontal task structuring, centralization versus decentralization of decisionmaking, and corporate sanctions of employees for negligence. Therefore, they argue for increased integration of environmental considerations throughout these corporate management incentive systems.

Hart (1995) proposes that proactive environmental management is itself potentially a strategic resource that can produce competitive advantage, especially for firms whose effectiveness in socially complex skills (such as total quality environmental management commitments, continuous improvement, cross-functional management, and interactions with the public) allows them to achieve greater economic advantages from pollution prevention, product stewardship, and sustainable development. Russo and Fouts (1997) concur, having examined 243 firms over two years, and conclude empirically that environmental performance and economic performance are positively linked, with the returns on environmental performance being higher in high-growth industries.

Like Hart and others, Florida and others (1999) argue that internal organizational factors, not only external pressures, play a fundamental role in the ability of business organizations to adopt advanced environmental practices. On the basis of a structured field research study that involved more than 100 interviews at matched pairs of 11 facilities in several industries, they conclude that organizational resources, particularly specialized environmental resources, provide the embedded capacity that allows sample facilities to implement environmental innovations. They also find that organizational monitoring systems play a crucial role in the adoption of environmentally conscious manufacturing practices. Finally, they find that such organizational resources tend to operate best as a system, creating the capacity to respond to internal opportunities and external events.

Klassen and Whybark (1999) investigate more closely the differences in performance associated with investments in pollution prevention, pollution control, and management systems. They conclude, theoretically and with empirical confirmation, that investments in prevention produce improvements in manufacturing and environmental performance, whereas investments in pollution control merely move pollutants around among different environmental media while adding costs and worsening manufacturing performance. Even proactive environmental policies provide little competitive advantage by themselves: what matters to economic competitiveness and to environmental performance is developing the capability to deploy pollution prevention technologies effectively. These findings concur with earlier empirical work by Hart and Ahuja (1996), who find that pollution prevention and emission reductions have a positive effect on industrial performance.

What remains to be studied in greater detail, Klassen and Whybark note, is whether allocating resources to management systems is a precursor to developing strategic organizational resources that favor the effective implementation of pollution prevention technologies (Hart 1995; Russo and Fouts 1997). To invest most effectively in pollution prevention, they argue, firms must develop strategic organizational resources to enable the recognition and deployment of pollution prevention technologies at the plant level, and then must ensure that plant-level personnel are given the latitude and the incentives to apply these capabilities to environmental issues in manufacturing, regardless of any corporate environmental policy. EMSs offer a potential organizational resource for this purpose, they suggest, but one that is not yet clearly proven.

Other researchers' findings also underscore the importance of determining how far and how fast the environmental management practices of leading firms are spreading to others. Arthur D. Little Inc. surveyed 185 North American EHS executives and found widespread barriers still blocking integration of EHS into the mainstream of their corporations' business, particularly in their continuing difficulty in persuading management that EHS is a legitimate core business issue (Arthur D. Little Inc. 1995; Meima 1997). Florida (1996) surveyed 212 corporate leaders on their environmental strategies and manufacturing practices in 1995, and at least at that time, the key drivers they perceived were still regulations and corporate citizenship; factors related to the desires to improve technologies and productivity and to serve key customers followed; and such factors as competition, markets for green products, and pressure from environmental organizations trailed more distantly.

Clinton Andrews (1998) also surveyed 116 Fortune 500 corporate leaders' perceptions of environmental business strategy. He finds that although most advocate environmental protection as a social goal and a core business objective, they still nonetheless perceive environmental considerations as more associated with risk and cost reduction than with value-adding competitiveness objectives such as productivity, prices, and sales. He also finds that in the firm's environmental decisions, respondents perceive environmental decisionmaking as still dominated by legal requirements, company values, and public perceptions overall. However, larger firms focus more on competitors' actions, company values, public perceptions, industry norms, and exhortation by public figures than do smaller ones, and multinational firms pay more attention to scientific evidence, suppliers' actions, company values, industry norms, and exhortation than do domestic ones.

These studies of the influence of individuals' attitudes on environmental management suggest the importance of an additional theoretical approach, which emphasizes not the business's strategic resources per se but individuals' perceptions of them. Meima (1997) proposes that the roles played by individuals throughout the organization and their interactions are an essential consideration and that they are closely intertwined with the peculiar nature of environmental issues, which have an "ecological logic" that is at first alien to traditional managerial discourse and therefore must be legitimized. Meima proposes using the "sense-making" tradition of organizational theory (Weick 1995) as an approach to this problem, investigating the ways in which individuals in the organization make plausible sense of environmental considerations and integrate them into their roles and interactions in ways that foster the development of the social complexity assets advocated by Hart (1995).

Research on EMS adoption itself, although still quite limited and largely atheoretical, offers some grounds for optimism. Stenzel (2000) notes that ISO 14001 was developed by deliberations among large transnational corporations with four principal motives: to promote sustainable development; to harmonize standards and procedures worldwide; to promote a new paradigm of self-management as an alternative to traditional government regulation; and to forestall further government regulation, especially at the international level.[4] Skeptics also criticize the ISO 14001 model for its origins in a relatively closed, self-appointed business organization; for the absence of any binding linkage to environmental performance standards, even regulatory compliance, other than those self-selected by the firm; for the absence of any requirement for public reporting and disclosure; and for its reliance on self-enforcement and on the standards and qualifications—as yet not clearly demonstrated—of third-party certifiers (Stenzel 2000). Some of the best-informed environmental observers in nongovernmental organizations conclude that EMSs can further principles of sustainability, help regulatory agencies achieve policy objectives, and improve relationships among stakeholders. In addition, they believe that the ISO 14000 standards can play a positive role in the greening of global commerce but cannot alone satisfy public policy objectives and, in particular, need to incorporate a meaningful public reporting requirement (Morrison and others 2000).

Early discussions of ISO 14001 certification assumed that because of its cost and effort burdens, the standard would be of interest mainly if not exclusively to large transnational corporations, such as those that initially negotiated the standard. Hillary (1999), however, reports on a meta-analysis of 33 studies of EMS implementation by small- and medium-sized enterprises (SMEs), primarily in Europe and the United Kingdom. Overall, SMEs that adopted EMSs find real and valuable benefits from doing so.[5] However, Hillary also identifies significant barriers to adoption. Internal barriers to EMS adoption are more important than external ones, particularly the scarcity of human (rather than financial) resources, practical problems with determining environmental aspects and assigning significance, the interruptibility of the process in an SME setting, lack of knowledge about EMSs and their potential benefits, and attitudes that the environment simply was not a core SME business issue or one that offered economic benefits. Customers are the key drivers for the adoption of EMSs by SMEs and have influence far beyond that of any of the other stakeholders cited; however, legislation and regulators are more important drivers for general environmental improvements in SMEs than are customers. Finally, implementation often requires more resources than expected, identification of noncompliance can be either a benefit (if the company could readily rectify the cause) or a problem (if it cannot or will not), and benefits to SMEs often have not materialized as expected (Hillary 1999).

Rondinelli and Vastag (2000) report that even in a firm with environmentally efficient operations already in place, ISO 14001 certification can have important behavioral and managerial impacts that contribute to better environmental performance, and these findings are reinforced by observations at other firms' ISO-certified facilities. Darnall and others (2001) report similar results, showing that several firms that had mature EMSs in place before adopting ISO 14001 still reported experiencing substantial benefits as a result of improved organizational control, communications, and manufacturing efficiency, all of which improved environmental performance.

Finally, the emerging business literature on corporate environmental management promotes a presumption that industry itself now best understands what drivers are most appropriate. It suggests that government incentives should therefore focus on rewarding voluntary business efforts to deliver environmental performance superior to the requirements mandated by statutes and regulations and on eliminating perverse incentives caused by some environmentally damaging taxes and other policies (for example, Schmidheiny 1992; Smart 1992; Hawken and others 1999; note that traditional environmental economists also would endorse at least the latter prescription).

An important issue for research, however, is the sustainability of such EMS commitments themselves, over time and across changes in personnel and in organizational ownership, structure, and management. In the context of the resource-based theory of the firm, Russo and Fouts (1997) note that industry transitions may render previously critical resources only marginally valuable. If industrial society evolves to the point where sustainable development is the norm, as Hart (1995) suggests, then technological, organizational, and human resources that serve a firm's environmental aims now should be even more valuable. But Richard Andrews (1998) also notes that real externalities and commons problems nonetheless continue to exist and that others have been ameliorated only by the existence of costly regulations in the shadow of which businesses now calculate strategic resources. The emerging theoretical arguments of a business case for EMSs—and for environmental performance improvements more generally—therefore must still be systematically and empirically proven against the impersonal forces of price and profitability in product and investment markets. They also must be shown to be generalizable not only for self-selected leading firms but also for all businesses that have a significant impact on the environment.

The NDEMS Database and Research Program

The most important question for EMS research is arguably the most obvious: what effect does the implementation of an EMS have on the environmental and economic performance of an organization that adopts it? This question is centrally important not only to businesses but also to regulatory officials, environmental groups, and affected communities.

Other questions also are interesting and important. For example, what motivates organizations to implement and certify EMSs? What do they expect to gain from them, and what do they actually gain? Is an EMS valuable or even necessary to compete in international markets? Is it important to major customers or suppliers, and if so, why? Does it make a difference to investors, insurers, and other important stakeholders? Does it help the organization in other ways, such as improved coordination among managers and divisions or greater worker involvement? Does it in fact reveal new opportunities for cost-effective pollution prevention, for reducing regulatory costs, or for more efficient business practices? Who actually is involved in adopting and implementing an EMS—within and outside of the facility— what difference does it make how and by whom it is carried out? Finally, why have even some nonmarket organizations (municipalities, state agencies, and federal facilities) decided to adopt such systems, and what have they gained from doing so?

One can ask equally important questions about the value of EMSs for achieving public policy goals. Does the implementation of an EMS improve businesses' environmental performance and reduce environmental impacts? Does it improve regulatory compliance, or does it at least improve self-monitoring, so as to reduce the taxpayer costs of monitoring and enforcement? What changes in federal and state regulations should be considered, if any, to promote EMS innovations that benefit public environmental goals, and to avoid any undermining of those goals? Does the EMS process improve relationships between businesses and their neighbors and communities, and between businesses and environmental and other citizen organizations? Does it improve environmental performance by suppliers and customers as well as by the primary business itself? What difference, if any, does third-party certification make?

NDEMS was developed to provide a basis for answering such questions. A joint initiative of the University of North Carolina at Chapel Hill and the Environmental Law Institute, NDEMS is supported by EPA in cooperation with MSWG, 10 state environmental agencies,[6] and, so far, more than 50 businesses and other organizations that have agreed to contribute data.

NDEMS was designed to include data on EMS implementation from pilot facilities that received assistance from state or federal agencies, plus nonpilot comparison facilities, using identical data collection protocols for each. The goal was to determine the effects of ISO 14001 and similar EMSs on six kinds of outcomes: environmental performance, environmental conditions, economic performance, regulatory compliance, pollution prevention, and engagement with stakeholders.

The design of the study is a longitudinal comparative case analysis in real time. It is specifically designed to collect facility-level data, because such data are necessary to examine actual changes in environmental performance and are the building blocks out of which any broader generalizations about corporate environmental performance must be constructed. For each facility, the research team administers a baseline protocol that captures three years' retrospective data to establish the environmental performance levels before EMS implementation; an EMS design protocol, which elicits data on the EMS implementation process as well as its substantive content (for example, specific environmental aspects, impacts, objectives, and targets); and a series of update protocols, to be administered annually over the following two to three years, to capture subsequent changes in environmental, economic, and other outcomes as well as refinements to the EMS itself. All data are subject to careful quality control procedures, including rechecking with the facilities before final inclusion in the database, to avoid errors or misinterpretation.

As of June 2000, baseline data were complete for more than 50 participating facilities, and initial EMS design data had been collected for more than 30 facilities. EMS design data were scheduled to be completed by the end of 2000, and update data on changes in performance were scheduled to be collected during 2001 and 2002. The protocol data were being augmented by on-site case studies for selected facilities.

Preliminary Results

What Kinds of Facilities Are Adopting Certified EMSs?

Baseline data analysis reveals several interesting characteristics of the NDEMS facilities. In contrast to early assumptions that EMSs would be adopted and certified only by large transnational corporations, EMSs are being implemented by facilities of all sizes and in many sectors. The database includes private- and public-sector facilities, large and small businesses, and simple and complex operations. Of the more than 50 facilities in the database so far, only 8 (16%) had more than 1,000 employees; 30 had 100–999 employees, and 12 employed fewer than 100 employees. Whereas 34 facilities were part of a larger parent organization, 16 were not; 76% did business overseas, and 24% did not. More than a dozen sectors of the economy were represented, including chemicals, electronics, food processing, machinery, metals, pharmaceuticals, pulp and paper, printing, transportation, and utilities; seven facilities were operated by federal, state, or local governments.

Second, a sizable majority of the facilities (88%) had some prior experience with environmental or nonenvironmental management systems, and nearly two-thirds reported participation in other voluntary environmental management incentive programs.[7] These findings tend to support Florida's contention (see Chapter 4 by Florida and Davison) that innovative firms are likely to be innovative across multiple dimensions.

However, the facilities were not idiosyncratically "clean." Most were regulated under air, water, and/or hazardous waste statutes, and more than 60% generated TRI-reportable quantities of toxic pollutants. A dozen major violations had occurred at 3 of these facilities, and 78 minor violations at 17 of them; most of them represented violations of emission or discharge limits or violations of monitoring requirements. Although 48 of these violations were self-discovered, 36 were discovered by regulatory inspectors; 58 were discovered within one day, but 16 only after more than two months. Similar issues were reported for "noncompliances" that were not cited as formal violations: 19 facilities experienced 27 actual noncompliances, and 25 facilities experienced 53 potential noncompliances, most frequently involving either emissions or discharges that exceeded permit limits, or unauthorized releases of other pollutants. Of these noncompliances, 52 were self-discovered, 10 by formal facility audits; only 3 were discovered by regulatory inspectors. Whereas 50 were discovered within one day, 21 were discovered only after more than two months. All these results suggest potential benefits from more systematic environmental management procedures.

Forty-three of the facilities reported that they already involved some interested parties in environmental management decisions during the three baseline years, most frequently nonmanagement employees, owners, and shareholders. About half involved local government agencies, but less than a dozen included environmental or other local citizen groups, community advisory groups, or neighbors. However, 25 facilities reported that they planned to institute or expand procedures for involving interested parties in their decisions.

Finally, the baseline data suggest clear performance differences between facilities that did and did not have a formal pollution prevention plan in place. Table 2-1 shows that facilities that had such plans were far more likely to involve suppliers and customers in pollution prevention initiatives, consider pollution prevention in product design and business planning, use materials accounting, have pollution prevention teams and training, and reward their employees for pollution prevention initiatives. These differences support Klassen and Whybark's (1999) speculations concerning the possible

Table 2-1. Association of Pollution Prevention (P2) Plans with Facilities' P2 Activities

	Facilities (%)	
Activity	With a P2 Plan (n = 23)	Without a P2 Plan (n = 19)
Involves suppliers in P2	78	42
Involves customers in P2	52	42
Incorporates P2 in product design	61	42
Incorporates P2 in business planning	57	26
Uses materials accounting	78	47
Uses P2 teams	57	32
Provides P2 training	70	26
Rewards employees for P2	48	26

Note: n = total number of facilities.

value of management systems for promoting pollution prevention. They also suggest the potential for similar differences between facilities that do and do not implement formal EMSs.

What Motivates Facilities to Adopt a Certified EMS?

Analysis of preliminary data on EMS content and design processes also suggests interesting and useful findings. On the basis of data for 31 of the 50 facilities, we examined responses to the question of what factors had greatest importance in their decision to adopt a formal EMS. Possible answers included a parent company requirement for EMS adoption; pressure from regulators, customers, shareholders, or others; expectation of benefits, such as increased revenues, reduced costs, better insurance rates, regulatory benefits, competitive advantage, or value as a marketing or public relations tool; a desire to improve environmental performance or to fulfill the facility's own environmental principles; a desire to improve employees' participation in environmental management; a desire to improve regulatory compliance; or the fact that government assistance made EMS adoption attractive.

Each possible motivation was ranked high, medium, low, or not applicable by each responding facility; the results are highlighted in Figure 2-1 for the percentage of facilities that ranked each factor as of "high" or "medium" importance. Overall, improved environmental performance and consistency with the organization's principles were the two strongest motivating factors, with compliance and employee participation next; cost reduction, regulatory benefits, and competitive advantage also were considered important by

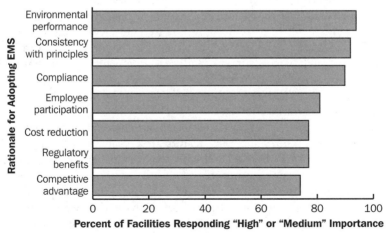

Figure 2-1. Seven Highest Motivations for EMS Adoption (31 Facilities)

Note: A small number of facilities did not respond in some of these categories.

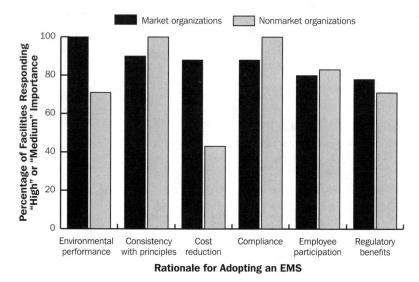

Figure 2-2. Differences in Motivation for EMS Adoption by Market vs. Nonmarket Organizations (31 Facilities)

Note: A small number of facilities did not respond in some of these categories.

approximately three-quarters of the facilities, and enhancing employee participation by more than two-thirds; all other factors ranked considerably lower.

A second question we asked was whether market and nonmarket (business and government) facilities have different motivations for EMS adoption. The results (presented in Figure 2-2) reveal some clear differences. For businesses, improving environmental performance was clearly the motivation most frequently asserted as important, followed closely by consistency with principles, cost reduction, and compliance improvement. For nonmarket organizations such as government facilities, in contrast, consistency with principles and compliance were the most important motivations. Cost reduction was considered important by only 43% of the nonmarket organizations, far less than for businesses.

Third, we examined whether customer and/or shareholder pressures were important factors in EMS adoption; results are presented in Figure 2-3. For all facilities, these influences were less important than those shown in Figure 2-1, but they varied significantly by size of facility and by whether the facility participated in foreign markets. Large facilities most frequently cited customer pressures as important, but interestingly, they were more often concerned about pressures from domestic customers than from international customers or shareholders. On the other hand, medium-sized facilities were less concerned about such pressures in general but were slightly more concerned about international than domestic customers; the small facilities

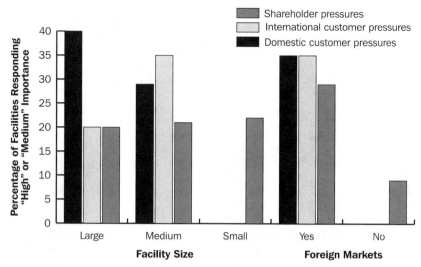

Figure 2-3. Importance of Customer and Shareholder Pressures in EMS Adoption (31 Facilities)

Note: A small number of facilities did not respond in some of these categories.

were concerned about shareholders but not at all about domestic or international customers. Interestingly, concern about both kinds of customers was centered in facilities that were active in foreign trade; for those that were not, neither type of customer pressure was perceived to be important.

Finally, we analyzed the importance of government assistance to different sizes and types of facilities; results are shown in Figure 2-4. Clearly, government assistance was perceived as very important by small businesses and by government (nonmarket) organizations and distinctly more important by facilities that were not active in international trade and by those that were not part of a larger organization. These data tend to confirm the importance of federal and state pilot assistance programs in helping these kinds of organizations to develop EMSs.

Other Aspects of EMS Design

We have done more detailed analysis on EMS design and implementation by 18 facilities whose initial data were relatively complete, representing eight industrial sectors in nine states. These first impressions suggest potentially interesting findings if they prove to be consistent across the full set of NDEMS facilities.

First, the responses appear to show that whereas not all facilities reported direct economic net benefits from EMS adoption, most believed that it had been a worthwhile process. Several explicitly stated that it had been suffi-

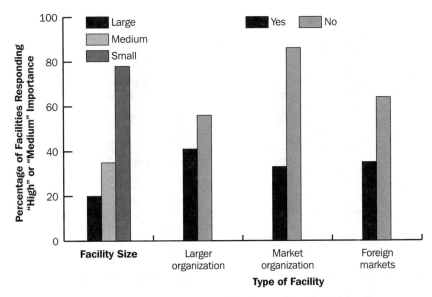

Figure 2-4. Importance of Government Assistance in EMS Adoption, by Facility Size and Type (31 Facilities)

Note: A small number of facilities did not respond in some of these categories.

ciently valuable that they would do it again, even though it might not pay for itself on any strict economic basis.

Second, nearly all 18 facilities used the EMS design process as an opportunity to investigate thoroughly all activities and areas of their facilities and to identify those that would have a potential impact on the environment. The few exceptions were facilities that perhaps relied too heavily on readily available, generic aspect and impact checklists rather than designing a specific process for their facility and thus bypassed part of the critical thought process of identifying their own distinctive aspects and impacts.

Third, the facilities used various procedures to identify environmental aspects and impacts and to determine their significance. Of the 18 facilities, five used group brainstorming to develop an initial list of environmental aspects and impacts; nine facilities relied on the environmental staff to compile a list of activities, and from it they derived a list of environmental aspects and impacts; three facilities asked each department to contribute specific lists of activities relative to their work and to determine the activities' associated environmental aspects and impacts; and one facility invited all employees to contribute such lists.

After lists of aspects and impacts had been compiled, 11 of the facilities created formal rating systems to evaluate their significance. Most of these used a two-tiered scoring method, based first on a combination of severity, probability, and duration or frequency of impact and second on a combination of compliance, legal or regulatory concerns, community concerns, and business or technical feasibility concerns. The facilities then combined the two sets of scores in various ways to come up with an overall significance score. Five facilities used group brainstorming or managerial judgements rather than formal rating schemes, and three facilities did not provide information on the techniques they used.

Determining significance was not always a straightforward task, even when formal systems were used. Three of the facilities specified a certain score as significant at the outset. Four facilities categorized the top 10 scores or a percentage of the top scores as significant. However, seven others weighted legal, regulatory, or compliance scores much higher than other considerations, ensuring that aspects and impacts that had a legal, regulatory, or compliance component would be judged significant.

Fourth, all 18 facilities' EMS design teams were headed by the facility environmental manager, and most were composed primarily of other environmental and engineering staff. These teams generally did not include non-management or hourly employees but a group of mid- to upper-level managers, usually including the plant manager, the facility engineering manager, and the maintenance manager. In six cases, the team was made up of all members of the facility's senior management team. In 8 of the 18 facilities, representatives from all departments were involved. Hourly employees were formally involved at only three facilities: in each of these cases, department managers solicited input from them directly in identifying activities within their departments that had potential environmental impacts. However, those facilities that did involve a wide variety of employees reported significant additional benefits from the process, namely, a heightened and more widely shared awareness of environmental issues among employees and a shared vision for addressing them.

Consultants were included as members of two facilities' EMS development teams, although many facilities sought the advice of consultants throughout the EMS design process. State government technical assistance staff were involved at six facilities in three states.

External stakeholder groups were involved in only two facilities' EMS development processes. At one facility, an external stakeholder group was established at the outset of the EMS design process, before the environmental policy had been articulated. At the other facility, external stakeholder input was incorporated when objectives and targets were being set.

Finally, more than half of these initial 18 EMSs had just been developed during the past year as part of a state pilot project, and most of them had set only a few short-term objectives and targets focused on compliance and/or pollution prevention.[8] In contrast, four of the facilities—those that had already prepared EMSs on their own and had implemented them for at least three years—exhibited objectives and targets that were more varied and more integral to the facilities' long-term environmental plans.[9] For example, one facility with a preexisting ISO 14001–certified EMS had explicitly incorporated principles of environmental sustainability into its EMS and, in so doing, had shifted its emphasis from short-term compliance improvements to long-term product stewardship. It will be important to observe whether the newly initiated EMSs of pilot facilities evolve in this way over time or whether state assistance proves to have been a structurally biasing incentive in favor of emphasizing short-term compliance improvement over other potential EMS priorities.

Further Research Plans

As the remaining EMS design data are added to the database, UNC–Chapel Hill and ELI are conducting research analyses on several issues that are of immediate interest and importance to public policy, business, and environmental stakeholder groups. They include five main issues.

First, what are the most interesting similarities and differences among the EMSs themselves? How do they differ in scope—facility-wide, or merely selected operations? How do they differ in priorities—improving compliance, improving regulated performance beyond compliance, or improving nonregulated aspects of environmental performance? How do they differ in objectives and targets—how far and how fast do they choose to push themselves to improve? Are there patterns of difference by sectors, facility size, public versus private ownership, or other factors? And finally, is the EMS systematic and strategic in the objectives and targets it recommends, or merely incremental and ad hoc?

Second, what difference does the EMS design and implementation process make? Who decides to establish an EMS, and what the process and its goals will be, and with what expectations for it? Who participates: only the EHS office; a small but broad core team; or other employees, consultants, outside stakeholders and community representatives, state technical assistance staff, third-party auditors, and others? What differences in the EMS content and outcomes result from differences in the process by which it is created?

Third, what costs and benefits do the staff of participating facilities perceive from EMS implementation so far? How well are the costs and benefits tracked, and how comprehensively are they documented? Were any benefits or costs unexpected? Do participants generally agree on these perceptions, or are there important differences? Do they consider the costs worthwhile and justifiable in relation to the benefits gained? Why or why not?

Fourth, what differences do state or federal pilot programs make to these outcomes? Are state assistance and incentives important factors in EMS content and outcome, and if so, how do they change the results? Are these programs producing benefits commensurate with the special allocation of staff effort and costs to them? What kinds of benefits are produced—better environmental performance, reduced state monitoring and enforcement costs, improvement of regulatory processes, improvement of regulatory relationships, or others?

Fifth and finally, what motivates facilities to adopt a formal EMS, given the considerable cost and effort necessary to do so? Who makes such decisions, what factors lead them to do so, and are these motivations similar or different across different sizes and characteristics of facilities—for instance, market businesses versus government agencies, facilities that are active or are not active in foreign trade, facilities that are or are not parts of larger organizations, and other characteristics?

Beginning in 2001, after the update data are collected on post-EMS adoption performance, we shall look at four additional issues:

- What differences in actual results from an EMS can be documented— changes in environmental and economic performance; regulatory compliance; pollution prevention measures; and relations with workers, communities, and other stakeholder groups? For instance, what additional benefits and costs result from involving interested outside stakeholders in the EMS process, and how does the involvement of inside and outside parties change as a result of adopting an EMS?
- What are the benefits and the costs of an EMS to the facility, to government, and to the public?
- What differences do state assistance and third-party certification make?
- What evidence is there of continuing environmental improvement over time or, alternatively, of any slackening of commitment that might occur after the initial implementation process?

Longer-term research also is needed concerning the stability or evolution of EMS goals and commitments over time, particularly through changes in personnel and in organizational ownership and control. A stated commitment of EMS adoption is to continuous improvement in environmental performance. However, it also is possible that such commitments would survive neither the replacement of the individuals who made and implemented the original commitments nor changes in competitive pressures in either product or investment market conditions, let alone the changes in priorities and internal organization that often accompany a corporate takeover or buyout (or, in the case of a public-sector facility, a change in elected political authorities). These issues need careful and ongoing study if EMSs are to be trusted as a "voluntary" approach to achieving public environmental goals.

Limitations

The NDEMS database has valuable potential for investigating many sorts of questions concerning EMS implementation. However, its limitations should be noted.

First, so far, the database consists of a heterogeneous set of more than 50 facilities, enough to document many important similarities and differences but not enough to produce statistically conclusive generalizations about entire industrial sectors. All the facilities are necessarily volunteers, which almost certainly implies an upward bias in the sample. That is, these facilities are proud enough of what they are doing that they are willing to share their data with us and to cooperate with their state environmental agencies.[10] On many questions, the data report the perceptions and assertions of individuals in each facility, albeit individuals responsible for the EMS implementation process and carefully quality-checked with them, not all of which can be independently verified with documentary evidence.

Second, in this kind of study, the research process and protocols themselves may influence EMS implementation in directions other than those the facility would have pursued on its own. For instance, to what extent does state technical assistance to pilot facilities influence them to focus more on compliance and pollution prevention than on unregulated aspects of their environmental performance (such as product stewardship or conservation of energy or water)? To what extent does even paying attention to our research protocols redirect their emphasis to the kinds of factors we are asking them about, at the expense of others? Some influence is probably unavoidable, but it is a particular challenge to our aspirations to draw conclusions from these data that might be applicable to other facilities implementing EMSs on their own.

Third, some of the data could perhaps provide clues to confidential business information about competitive processes. Participating facilities have been extremely generous about sharing data with us, but in at least a few cases they have found it necessary to withhold specific data to protect confidential business information.

Fourth, facility-level data alone do not answer all important questions about the value and effectiveness of EMSs. The questions we are investigating in these facilities certainly should be replicated for additional comparison groups of facilities and augmented with more detailed on-site case studies. They also should be replicated for facilities in other countries, to compare national and cultural differences in the uses of these procedures. Some of these facilities could well be different locations of the same parent corporations; others should be facilities that do not share that common influence and that might therefore reveal important differences in processes and outcomes rooted in different national jurisdictions, economic systems, and cultures.

Additional questions should be studied concerning corporate-level use of EMSs. For instance, are EMSs used by firms to make strategic decisions? If so, how? Are there strategic motivations for introducing consistent types of EMSs throughout an entire corporate structure, or even to its suppliers and/or customers? What is the impact of EMS implementation on customers? On suppliers? Do such initiatives facilitate additional or different benefits from those available at the facility level alone? Examples might include changes in corporate-level cost accounting systems to allocate environmental costs more explicitly to specific activities that generate them (activity-based costing), which could not be altered at the facility level alone, or changes in the strategic configuration of an entire firm to achieve pollution prevention efficiencies among wastes from some divisions and inputs to others ("industrial ecology").

Yet another set of questions concerns the process of third-party auditing and registration. What is the competence of the providers of these services? What standards and criteria do they use to support or withhold certification? How consistent are these criteria across registration providers? What are the practical incentives to these firms to apply stringent or lenient standards for registration, and what are the resulting dynamics of the third-party registration services industry over time?[11]

Finally, important public policy questions need to be evaluated in the context of EMS implementation. For instance, what sorts of regulatory flexibility might be appropriate in the context of an effective EMS, and with what conditions? How should agencies judge whether an EMS qualifies a facility for regulatory flexibility? How do EMSs fit into the broader environmental policy debate over requirements for scientific and economic justification of regulatory policy? And do government, business, and interest groups act and interact more or less productively in EMS implementation processes than they do in regulatory proceedings? What lessons does this offer for policy and procedural improvements?

All these questions offer promising and timely opportunities for research on EMS-related phenomena. The NDEMS provides a valuable starting point for many of them, and we welcome interaction with other researchers interested in using it and with those studying related questions.

Conclusion

The widespread introduction of formal EMSs into the practices of businesses that affect the environment offers a unique opportunity to observe the processes and the environmental and economic consequences of these initiatives and to compare similarities and differences across different firms, sectors, sizes, and other characteristics. From a public policy perspective, it also offers an unusual opportunity to look at the achievement of environmental and economic objectives through the eyes of the businesses whose actions are critical to those outcomes, rather than merely through the perspective of government agencies themselves. At the same time, it also should shed light directly on environmental policy questions such as the practical issues involved in improving regulatory compliance, environmental performance, and cost-effectiveness in monitoring and reporting.

Understanding the relationship between EMS adoption and actual environmental performance is critically important to future environmental initiatives at the federal and state level, voluntary and mandated. Moreover, what matters is not merely the fact of EMS adoption but the likelihood that key elements of some EMSs are more likely associated with superior environmental performance than EMS adoption per se. Understanding the motivations that contribute to a facility's decision to voluntarily reduce its environmental impacts, regulated and nonregulated, also is critically important. If the NDEMS data turn out to show that well-designed EMSs do contribute to achieving superior environmental performance, then government officials might appropriately consider policy changes to encourage the wider introduction and certification of EMSs that contain elements that promote superior results and, more important, to facilitate more effective and less costly means of achieving high environmental performance opportunities that EMSs may identify.

Acknowledgements

We acknowledge with gratitude the contributions of John Villani (formerly of the University of North Carolina at Chapel Hill) and the generous cooperation of the participating facilities and state project managers.

Notes

[1]With the exception of information that would reveal the identity of specific facilities, all National Database on Environmental Management Systems documentation is publicly available online at http://www.eli.org/isopilots.htm.

[2]For instance, Ehrenfeld and Howard (1996) observed that overwhelmingly the principal organized activity of U.S. industrial trade associations is lobbying to restrict or control environmental regulation rather than to achieve better environmental performance. They also noted that even many leading "green-image" firms, such as AT&T, said little or nothing about their environmental initiatives in their corporate annual reports, suggesting by implication that at least at that time, these matters were still considered inconsequential to the core of their economic and investment decisions.

[3]One major pharmaceuticals firm, for instance, responded to several major environmental embarrassments by establishing a corporate commitment to superior environmental performance by all its production units and built a widely respected corporate environment, health, and safety (EHS) staff to implement this commitment throughout its operations. However, it was subsequently acquired by another firm whose management policy was that each production unit should have decentralized responsibility for its own decisions. The corporate EHS staff was severely downsized, and most reportedly left the company.

[4]Note that ISO 14001 offered a worldwide alternative to two more stringent standards then being introduced in Europe: England's BS 7750 and the European Union's Eco-Management and Auditing Scheme (EMAS).

[5]Examples included particularly the attraction of new businesses and customers, satisfaction of customer requirements, improved environmental performance, assured legal compliance, and material and energy efficiencies as well as organizational improvements and efficiencies, financial savings, and broader attitudinal and communication benefits.

[6]The states are Arizona, California, Illinois, Indiana, New Hampshire, North Carolina, Oregon, Pennsylvania, Vermont, and Wisconsin.

[7]Examples of prior environmental management system (EMS) experience included some form of preexisting EMS, waste minimization planning, pollution prevention planning, compliance audits, internal and (less frequently) public environmental reporting, environmental best practices, and risk assessment or environmental accounting systems. Examples of prior nonenvironmental management systems included ISO 9000 certification, total quality management, materials accounting systems, just-in-time inventory systems, and the Occupational Safety and Health Administration voluntary protection program. Voluntary environmental incentive programs included state voluntary environmental management programs and EPA's 33/50 or Green Lights programs; others (one or two each) mentioned were the Charter for Sustainable Development, CERES Principles, Business for Social Responsibility, and EPA's Green Star Program.

[8]Interestingly, one facility even included objectives and targets that had already been reached before the EMS was complete—perhaps to use early and easy successes to build momentum for further implementation, or perhaps simply to use the EMS document for good public relations.

[9]Of the 18, two addressed product stewardship, two others included the development of employee environmental awareness programs as specific objectives and targets, and one incorporated an objective to design and implement an environmentally friendly cleaning program.

[10]Whether they were proud of their performance because it resulted from EMS adoption or simply because it represented high environmental performance due to management leadership more generally is an important distinction to consider. In fact, some participating states barred the participation of facilities that had a history of significant compliance problems; as a result, some facilities that might otherwise have shown more dramatic changes as a result of EMS introduction are not included in the study so far. One hope for the future is to obtain data from a comparison group of compliance-mandated EMS adopters. Given these issues, we have conducted careful baseline data collection over several prior years to distinguish more

carefully the facilities that were already high environmental achievers from those that were not.

[11]The National Academy of Public Administration is concurrently conducting such a study in 2000–2001, with support from the U.S. Environmental Protection Agency.

References

Allenby, Brad, and D. Richards (eds.). 1994. *The Greening of Industrial Ecosystems.* Washington, DC: National Academy Press.

Andrews, Clinton J. 1998. Environmental Business Strategy: Corporate Leaders' Perceptions. *Society and Natural Resources* 11: 531–540.

Andrews, Richard N.L. 1998. Environmental Regulation and Business "Self-Regulation." *Policy Sciences* 31(3): 177–197.

———. 1999. *Managing the Environment, Managing Ourselves: A History of American Environmental Policy.* New Haven, CT: Yale University Press.

Arthur D. Little Inc. 1995. *Results of Arthur D. Little's 1995 Survey on "Hitting the Green Wall."* Cambridge, MA: Arthur D. Little Inc.

Cairncross, Frances. 1991. *Costing the Earth.* Cambridge, MA: Harvard Business School Press.

Darnall, Nicole, Deborah Rigling Gallagher, and R.N.L. Andrews. 2001. ISO 14001: Greening Management Systems. In *Greener Manufacturing and Operations,* edited by Joseph Sarkis. Sheffield, U.K.: Greenleaf Publishing, Chapter 12.

Davies, Terry, and Jan Mazurek. 1996. *Industry Incentives for Environmental Improvement: Evaluation of U.S. Federal Incentives.* Washington, DC: Global Environmental Management Initiative.

Ehrenfeld, John, and Jennifer Howard. 1996. Setting Environmental Goals: The View from Industry. In *Linking Science and Technology to Society's Environmental Goals.* Washington, DC: National Research Council, National Academy Press, 281–325.

Fischer, Kurt, and Johan Schot (eds.). 1993. *Environmental Strategies for Industry.* Washington, DC: Island Press.

Florida, Richard. 1996. Lean and Green: The Move to Environmentally Conscious Manufacturing. *California Management Review* 39(1): 80–105.

Florida, Richard, Mark Atlas, and Matt Cline. 1999. What Makes Companies Green? Organizational Capabilities and the Adoption of Environmental Innovations. Unpublished paper. Pittsburgh, PA: Heinz School of Public Policy and Management, Carnegie Mellon University.

Gabel, H. Landis, and Bernard Sinclair-Desgagné. 1994. From Market Failure to Organisational Failure. *Business Strategy and the Environment* 3(2): 50–58.

Hart, Stuart L. 1995. A Natural Resource-Based View of the Firm. *Academy of Management Review* 20: 986–1014.

———. 1997. Beyond Greening: Strategies for a Sustainable World. *Harvard Business Review* 75(1): 66–76.

Hart, Stuart L., and Gautam Ahuja. 1996. Does It Pay to Be Green? *Business Strategy and the Environment* 5: 30–37.

Hawken, Paul. 1993. *The Ecology of Commerce*. New York: Harper.

Hawken, Paul, Amory Lovins, and L. Hunter Lovins. 1999. *Natural Capitalism: Creating the Next Industrial Revolution*. Boston, MA: Little, Brown.

Hemphill, T.A. 1995. Corporate Environmental Strategy: The Avalanche of Change since Bhopal. *Business and Society Review* 94: 68–72.

Hillary, Ruth. 1999. *Evaluation of Study Reports on the Barriers, Opportunities and Drivers for Small and Medium Sized Enterprises in the Adoption of Environmental Management Systems*. Report submitted to the Environment Directorate, Department of Trade and Industry, U.K., October 5, 1999.

Jaffe, Adam B., Steven R. Peterson, Paul R. Portney, and Robert N. Stavins. 1995. Environmental Regulation and the Competitiveness of U.S. Manufacturing: What Does the Evidence Tell Us? *Journal of Economic Literature* 30: 132–163.

Kapp, K. William. 1950. *The Social Costs of Business Enterprise*. New York: Schocken.

Klassen, Robert D., and Curtis P. McLaughlin. 1996. The Impact of Environmental Management on Firm Performance. *Management Science* 72(3): 2–7.

Klassen, Robert D., and D. Clay Whybark. 1999. The Impact of Environmental Technologies on Manufacturing Performance. *Academy of Management Journal* 42(6): 599–615.

Kneese, Allen V., and Blair T. Bower. 1979. *Environmental Quality and Residuals Management*. Washington, DC: Resources for the Future.

Meima, Ralph. 1997. The Challenge of Ecological Logic: Explaining Distinctive Organizational Phenomena in Corporate Environmental Management. In *Corporate Environmental Management 2: Culture and Organisations*, edited by Richard Welford. London, U.K.: Earthscan, Chapter 3, 26–56.

Morrison, Jason, Katherine Kao Cushing, Zöe Day, and Jerry Speir. 2000. *Managing a Better Environment: Opportunities and Obstacles for ISO 14001 in Public Policy and Commerce*. Oakland, CA: Pacific Institute for Studies in Development, Environment, and Security.

Palmer, Karen, Wallace Oates, and Paul Portney. 1995. Tightening Environmental Standards: The Benefit-Cost or the No-Cost Paradigm? *Journal of Economic Perspectives* 9(4): 119–132.

Porter, Michael E. 1991. America's Green Strategy. *Scientific American* 264(4): 168.

Porter, Michael E., and Claas van der Linde. 1995a. Green and Competitive: Ending the Stalemate. *Harvard Business Review* 73(5): 120–134.

———. 1995b. Toward a New Conception of the Environment-Competitiveness Relationship. *Journal of Economic Perspectives* 9(4): 97–118.

Rondinelli, Dennis A., and Gyula Vastag. 2000. Panacea, Common Sense, or Just a Label? The Values of ISO 14001 Environmental Management Systems. *Environmental Management Journal* 18(5): 499–510.

Royston, Michael. 1979. *Pollution Prevention Pays*. London, U.K.: Pergamon.

Russo, Michael V., and P.A. Fouts. 1997. A Resource-Based Perspective on Corporate Environmental Performance and Profitability. *Academy of Management Journal* 40: 534–559.

Sarokin, David J., Warren R. Muir, Catherine G. Miller, and Sebastian R. Sperber. 1985. *Cutting Chemical Wastes*. New York: INFORM.

Schmidheiny, Stephen. 1992. *Changing Course: A Global Business Perspective on Development and the Environment*. Cambridge, MA: MIT Press.

Schot, Johan, and Kurt Fischer. 1993. Conclusion: Research Needs and Policy Implications. In *Environmental Strategies for Industry: International Perspectives on Research Needs and Policy Implications*, edited by Kurt Fischer and Johan Schot. Washington, DC: Island Press, 369–373.

Smart, Bruce. 1992. *Beyond Compliance: A New Industry View of the Environment*. Washington, DC: World Resources Institute.

Stenzel, Paulette L. 2000. Can the ISO 14000 Series Environmental Management Standards Provide a Viable Alternative to Government Regulation? *American Business Law Journal* 37: 237–298.

U.S. EPA (Environmental Protection Agency). 1990. *Reducing Risk: Strategies for Environmental Protection*. Washington, DC: U.S. EPA, Science Advisory Board

Weick, K.E. 1995. *Sensemaking in Organizations*. Thousand Oaks, CA: Sage..

Weidenbaum, Murray. 1979. *The Future of Business Regulation*. New York: American Management Association.

3

Factors That Shape
EMS Outcomes
in Firms

Jennifer Nash and John R. Ehrenfeld

In recent years, government at the federal and state levels has begun to explore the potential of nonregulatory policy instruments. This exploration has included granting limited regulatory flexibility to firms or facilities that demonstrate environmental excellence. Environmental excellence is performance that exceeds some normal baseline. One criterion of excellence used by agencies is the adoption by the firm of a satisfactory environmental management system (EMS). The rationale behind this approach is that properly constructed and operated EMSs will in theory lead to superior environmental performance. But theory and practice often part company.

In this chapter, we examine the complex relationship between a firm's EMS and its environmental performance. We begin by defining what a formal EMS is and describing three distinct types. An EMS always exists within the context of a firm's individual culture, its strategic goals, and its competitive position within its industry. These factors shape how the formal EMS works in practice. Because many firms today are adopting formal EMSs largely in response to external demands, the decision to adopt the EMS is not necessarily linked to a decision to invest in environmental performance improvement.

Agencies want to be able to separate excellent firms from the norm and would like to use EMS adoption as a means to make this distinction. The difference between what a formal EMS is in theory and the way it works in practice poses a challenge for agencies. This problem corresponds to the typical principal–agent situation that has been intensely studied in the economics field (Holmstrom and Milgrom 1994). Framing the problem in this way helps to clarify what both agencies and firms require to use EMS adoption as the basis for entry into the excellent tier of companies.

Defining an EMS

EMSs are formal structures of rules and resources that managers adopt to establish organizational routines that help achieve corporate environmental goals. They are a subset of management systems in general.

A company, like any other organization, is made up of people whose actions routinely produce a set of intentional outcomes. A manager of a company has a vision of these outcomes. This vision may be expressed as a formal mission statement and written policies and may contain statements about the company's values (that is, what it holds important). Such formal texts constitute the rules of management systems. Firms always have such rules, but in many cases—particularly in small firms—they exist only as some shared set of values and vision in the cognitive background of the actors.

In addition to these rules, companies create authority structures and provide technological resources that empower personnel to perform effectively in a routine fashion. The authority structures are communicated through organizational charts and the delegation of specific responsibilities. Technological resources are tools and information systems used to produce the desired outcomes. They include devices such as production equipment, environmental control technology, and information technology. Most important, these resources include the personnel involved in delivering the outcomes.

Typology of EMSs

During the past decade, individual firms, trade associations, and standards organizations have developed different types of EMSs. Each type poses distinct challenges and opportunities to agencies that are interested in using an EMS as a basis for participation in an excellence program.

Firm-Structured EMSs. Since the mid-1980s, many firms have developed EMSs to establish rules and make available the resources deemed necessary to meet established goals. The following examples describe the experiences of two firms that developed and implemented EMSs they structured themselves. The systems were implemented more than a decade ago, giving us the perspective necessary to assess their impact over time.

The managers that developed the firm-structured EMSs described here set goals and developed an implementation plan, monitored progress, and met regularly to discuss problems. They assigned responsibility for meeting environmental targets and instituted new organizational structures to gather information and track progress. They implemented training programs and undertook corrective action when progress was not on track. These categories of activities are common to all EMSs.[1]

Polaroid Corporation. In 1987, Polaroid's board of directors established the goal of reducing chemical use and waste by 10% per year (per unit of production) for the five-year period ending in 1993. Polaroid's managers created a complex accounting system to measure progress toward this goal at each of the company's 23 manufacturing plants. Managers used this environmental accounting system to measure quantities and treatment methods of each of the 1,700 materials the company used. Individuals throughout the company were told to track their groups' consumption of these materials, complete quarterly reports, and participate in meetings to discuss progress. Polaroid began publishing an annual environmental performance report beginning in 1988 on the basis of information developed through its environmental accounting system (Nash and others 1992).

Reductions were achieved on or ahead of schedule during the first years of the program. In the third year, progress began to slow. Over the full five years, chemical use and waste were reduced 5% per unit per year, and the consumption of all other materials (cardboard, paper, and plastic) was reduced 7% per unit per year. Managers found it difficult to justify the costs of meeting the annual 10% reduction goal, especially in light of declining profits.

Since 1995, Polaroid's business strategy has changed. The company relies less on internal manufacturing of components for finished products. In recent years, the reduction goals and accounting system have been replaced by a less intensive environmental management approach that focuses more on compliance than the old system did.

The Robbins Company. In the late 1980s, The Robbins Company, a small metal finisher and plater operating in southeastern Massachusetts, found itself in violation of its water discharge permit. While other platers in the area that faced similar regulatory requirements opted to hook up their waste treatment facilities to the town sewage treatment plant (at considerable expense), Robbins' managers decided to set the goal of zero discharge and thereby eliminate their need for waste treatment entirely. Not knowing whether this goal was feasible, they searched for an engineering firm that could design a closed-loop system.

Working closely with the company's president, the environmental manager developed a management system that was based on continuous reductions in water use. It required careful monitoring of material flows as well as training and communication to foster rigorous adherence to new procedures. Each worker was trained to understand his or her responsibilities under the new system. The closed-loop system was implemented in 1988. The new technology has saved Robbins more than $100,000 annually, improved product quality, and earned accolades from employees and environmental advocates (Berube and others 1992). Although Robbins has modified the system since 1988, the goal of zero discharge is still being achieved.

Trade Association EMSs. In the United States, seven trade associations have developed codes of environmental management practices that fit our definition of an EMS (Nash 1999). The best-known example is the Responsible Care initiative of the American Chemistry Council (ACC; formerly known as the Chemical Manufacturers Association). The Responsible Care initiative is a set of guiding principles and management practices that govern environmental and safety aspects of chemical manufacturing, distribution, product stewardship, and community relations. ACC requires that members adopt Responsible Care as a condition of membership.

The Responsible Care initiative constitutes an EMS guideline. It differs from some other EMS guidelines in that it addresses the content of corporate environmental goals. This initiative was a direct attempt by U.S. chemical manufacturers to institutionalize new norms of behavior for participants. The guiding principles of Responsible Care call on firms to "operate ... facilities in a manner that protects the environment" and "make health, safety, the environment, and resource conservation critical considerations for all new and existing products and practices" (CMA 1998). ACC firms must establish programs to involve people who live near their facilities. The firms must monitor the environmental and safety practices of their distributors and customers and must discontinue business relationships with firms whose practices are inconsistent with Responsible Care. To work toward achieving these substantive goals, firms are required to develop policies, assign responsibility, and document progress in each of these areas—to implement more than a hundred management practices.

Members must audit their own Responsible Care progress and report the results to the trade association. In addition, members of the ACC can undergo a management systems verification. ACC has hired a private consultant, Verrico Associates, to handle this process. Verrico assembles a team of managers from peer companies and external stakeholders. Findings are generally divided into two categories: strengths and opportunities for improvement. The verification team visits sites and corporate offices and verifies that practices are consistent with Responsible Care. As of July 2000, about one-half of the ACC's members had undergone this review.

Trade associations in the chemicals distribution, petroleum, forestry, and textile industries also have adopted EMS guidelines for their members. Table 3-1 lists the mechanisms these trade associations have adopted to ensure that members abide by code requirements. The primary mechanism of control is peer pressure (Simmons and Wynne 1993). This method can only

Table 3-1. Authority Structures Established by Trade Association Codes

Organization and standard	Trade association organizes meetings to share best practices	Trade association compiles information on members' performance in aggregate	Trade association reveals individual firm's performance to board of directors	Third party verifies EMS is in place	History of trade association expulsion of members for noncompliance
American Chemistry Council: Responsible Care	X	X	X	Discretionary	
National Association of Chemical Distributors: Responsible Distribution Process	X	X		Required	X
American Petroleum Institute: Strategies for Today's Environmental Partnerships	X				
American Forest and Paper Association: Sustainable Forestry Initiative	X	X		Discretionary	X
American Textile Manufacturers Institute: Encouraging Environmental Excellence (E3)	X	X			

work if managers know how their firm's code performance compares with others in their industry. All of the trade associations that have promulgated codes hold periodic conferences to share best practices. Managers of firms whose activities are falling behind presumably feel pressure to improve. Recently, the American Chemistry Council board of directors made the important decision to rank some aspects of members' code performance, on a scale of 1 to 191 (the number of member companies), and distribute this ranking to its membership. Previously, the ACC, like other trade groups with codes, prepared code performance reports based on information aggregated from all members. Aggregated information is helpful in showing overall industry trends but does little to inspire laggards to improve.

Few empirical studies have documented how firms incorporate Responsible Care into their EMSs. In their 1995 survey of chemical firms' adoption practices, Howard and others (2000) found substantial variation in firms' implementation of Responsible Care requirements. Researchers identified four general categories of companies that had adopted the initiative: drifters, promoters, adopters, and leaders. *Drifters* were companies that said Responsible Care had little impact on their activities. Changes were limited to documenting existing practices. *Promoters*, who used Responsible Care mainly to promote a strong environmental reputation to external stakeholders, saw Responsible Care as an adjunct to existing environmental programs. It reinforced what they were already doing but did not cause them to rethink their activities. People in this group spoke of Responsible Care as "formalizing" and "standardizing" what they already did. *Adopters* were firms that saw Responsible Care as a valuable tool for improving their environmental practices. They credited Responsible Care with introducing new practices in community relations and distribution. Not only were environmental and communications staff handling Responsible Care activities; product managers, designers, and marketing staff also were involved. Finally, *leaders* spoke about Responsible Care being a "whole new way of thinking." They believed that whereas their environment, health, and safety (EHS) practices were strong before Responsible Care, the initiative offered a way to progress. In these firms, significant resources had been applied to Responsible Care implementation, and senior management took an active role in overseeing it.

All of the firms that participated in the study were at roughly the same stage of Responsible Care implementation, according to the self-assessments reported to ACC. What accounts for the differences in their responses? ACC explains that it expects members to implement practices "at the pace that is right for them" (CMA 1993). Members are not free to ignore Responsible Care, but implementation is a matter of individual firm responsibility (Rees 1997). Firms decide for themselves the level of resources that they will devote to Responsible Care and the specific activities they will undertake.

Standardized EMSs. In 1993, the International Organization for Standardization (ISO) organized a process to develop an international environmental management standard. Most of ISO's previous work had focused on product standards—precise specifications of product attributes. Developing a standard for EMSs required a new approach. Instead of offering detailed prescriptions of operating *practices*, ISO decided to work to define *procedures* firms must follow to identify environmental impacts, train workers, and document progress. The British Standards Institute (BSI) had used this approach in the draft EMS standard it issued in 1992. ISO committees followed BSI's lead and drafted an EMS standard they believed was flexible enough to be applicable to any organization (Uzumeri 1997).

ISO's work was strongly influenced by the theory of total quality management, which is based on the idea that greater quality (in the form of fewer defects, faster delivery time, and reduced operating costs) leads to lower costs to the producer and greater customer satisfaction. Under a total quality management framework, all activity flows from a business policy generated by top management. The primary purpose of the management system is to ensure that business activities are carried out in accordance with this policy. Managers periodically check to determine whether the organization is on track toward stated policy goals and to make any necessary corrections. Everything must be documented: the policies, the names of people responsible, the actions they will take to move the organization toward policy goals, and auditing and training procedures. Management systems based on total quality frameworks, like ISO 14001, emphasize management accountability to stated policy goals.

As BSI worked to finalize its EMS and international teams hammered out ISO 14001, the European Commission took steps to develop an EMS standard for the continent, known as the Eco-Management and Audit Scheme (EMAS). The three resulting EMS standards have much in common (Nash 1997). Organizations that participate in any of them must

- establish an environmental policy,
- identify environmental aspects of the facility,
- set environmental goals and objectives and establish programs,
- document procedures,
- assign responsibility,
- train workers,
- review progress, and
- continuously improve.

An additional component is the option that firms may hire third-party contractors to certify that their management systems comply with the standard. Third-party certification and associated registration is intended to ensure that all elements specified by the standard are in place, employees are aware of their roles and responsibilities, and the system is directed toward achieving the goals the organization has specified for itself in its policy and objectives. Only facilities that have had their systems reviewed and approved by a qualified third-party verifier may be registered to the standard.

ISO 14001 is relatively new, and experience in the United States is limited. Early indications suggest that individual firms will respond quite differently to its requirements and processes, as they have to those of Responsible Care. Not surprisingly, case studies of early adopters suggest that firms select goals (one of the initial steps to be taken under the ISO protocol) that are closely aligned with their current business objectives (Nash and others 1999). As the following examples suggest, firms' responses to the standard will likely vary according to their business needs, their regulatory environment, and the demands placed on them by customers and other stakeholders.

A large electronics manufacturer is using ISO 14001 to strengthen its environmental compliance program. For this firm, any legal transgression could jeopardize the U.S. Defense Department contracts that constitute the majority of its business. It specializes in the production of large, complex circuit boards. A single circuit board sells for as much as $200,000. Managers have decided not to include pollution prevention goals in their ISO-structured EMS. They would rather waste small quantities of chemicals than experiment with new processes that might yield lower quality.

In contrast, a large firm in the home and beauty care industry has adopted ISO 14001 to help workers identify opportunities for waste reduction. In this high-volume, low-profit business, managers have set the goal of exploring every opportunity to reduce resource consumption. Savings of a fraction of a cent per unit add up as products are manufactured again and again.

Comparison of EMS Requirements

The three kinds of EMSs we have described can be compared in terms of the kinds of environmental objectives they require firms to establish. Table 3-2 summarizes these objectives for trade association EMSs, ISO 14001, and the National Environmental Performance Track, an EPA-sponsored excellence program (discussed later, under Agency Response).

EMS guidelines promulgated by trade associations prescribe a common set of environmental objectives. These guidelines require members to involve communities in their environmental activities, assume stewardship over their products, prevent pollution, and improve their environmental management practices. ISO 14001 allows managers to set environmental objectives themselves. The standard requires only that the firm's policy include a commitment to regulatory compliance and continuous improvement in the EMS. Firm-structured EMSs reflect the goals and authority

Table 3-2. Comparison of Environmental Objectives of an EPA Excellence Program, ISO 14001, and Selected Trade Association Environmental Management Systems

Organization and standard	Explicitly complements regulations by requiring compliance	Requires continuous improvement in environmental performance	Requires product stewardship	Requires community involvement	Requires pollution prevention
U.S. EPA: National Environmental Performance Track (Achievement Track)	X	X		X	X
ISO: ISO 14001	X	X[a]			
American Chemistry Council: Responsible Care		X	X	X	X
National Association of Chemical Distributors: Responsible Distribution Process	X		X	X	X
American Forest and Paper Association: Sustainable Forestry Initiative	X	X	X	X	X

[a] There is disagreement over whether ISO 14001 actually requires environmental performance improvement. The standard defines continual improvement as "enhancing the environmental management system to achieve improvements in overall environmental performance in line with the organization's environmental policy."

structures of the individual firms that develop them. We do not attempt to draw general conclusions about their features.

EMSs also can be compared in terms of the mechanisms they require to ensure performance. Table 3-3 summarizes these features. The ISO 14001 system offers the option of third-party certification by people trained to provide objective confirmation to concerned outsiders that a firm has an EMS in place that is appropriate to its needs. Third-party certification also helps to provide external, nonregulatory oversight of environmental management. The EHS director of a large specialty chemical manufacturer explained, "I've found ... that people tend to let these management systems drift unless someone kicks you in the tail—that motivation by internal and external audits tends to keep the system current" (Switzer 1999, 91). The third parties used by trade associations are generally managers from other similar facilities. Although ISO 14001 registrars emphasize conformity with procedures, peer oversight tends to emphasize performance.

Environmental performance reporting is a feature of European EMSs but not those generally adopted by facilities in the United States. A central feature of EMAS is its requirement that companies publish statements of their significant environmental impacts, after validation by a third party. Even though many large U.S. corporations publish environmental reports and trade association EMSs encourage community involvement, disclosure of information to the public is generally not an EMS requirement in this country.

How do the features of EMSs affect the way environmental concerns are addressed by adopting firms? To explore this question, we turn to a discussion of the factors that motivate EMS adoption.

Motivations for EMS Adoption

Understanding a firm's motivations for adopting a formal EMS has become more complicated in recent years. External stakeholders, including customers and agencies, have begun to offer incentives to firms that adopt EMSs with specified features. As a result, discerning a firm's commitment to environmental performance improvement is difficult based on EMS adoption alone.

Adoption for Environmental Performance Improvement

Many companies adopt EMSs to strengthen their environmental performance. Such was the case for Polaroid and The Robbins Company. These firms implemented formal EMSs to improve regulatory compliance and reduce materials use and waste. Each faced significant pressure to improve its environmental performance from regulators and environmental advocacy

Table 3-3. Mechanisms to Ensure Performance Established by an EPA Excellence Program, ISO 14001, and Selected Trade Association Environmental Management Systems

Organization and standard	Uses peer pressure to ensure implementation	Offers the option of third-party verification	Requires publication of environmental performance report
U.S. EPA: National Environmental Performance Track (Achievement Track)			X
ISO: ISO 14001		X	
American Chemistry Council: Responsible Care	X	X	
National Association of Chemical Distributors: Responsible Distribution Process	X	X[a]	
American Forest and Paper Association: Sustainable Forestry Initiative	X	X	

[a] Verification required as condition of membership.

groups, and the programs implemented were a response to these pressures. The demands of these external stakeholders did not explicitly include formal EMS adoption, however. These companies used EMSs as tools to improve performance and thereby meet the expectations of external groups.

The "leaders" identified in the Responsible Care study were firms that were committed to environmental excellence and used Responsible Care to achieve it. It is too soon to assess the motivations of U.S. facilities that adopt ISO 14001. Early adopters may share characteristics with the Responsible Care leaders. A 1999 study of six ISO 14001 adopters in the chemical industry found that the primary motivation for ISO 14001 registration was the desire to improve compliance and environmental performance (Nash and others 1999).

Adoption in Response to Direct Demands

Many firms now associate EMS adoption with direct incentives from their trade association, customers, and agencies.

Peer Pressure. Many trade associations now require their members to adopt the trade association EMS guideline; they cannot belong to the association without doing so. Trade association EMS programs differ from the other kinds of EMSs discussed in this chapter because they represent collective action by firms in an industry. Although the requirements of trade association EMS programs are similar to those of ISO 14001, the motivations for joining are in many respects different. The ACC's stated purpose in establishing Responsible Care was twofold: to improve the chemical industry's environmental performance, and to communicate that improvement to "critical public audiences"—particularly regulators and community and environmental groups. Responsible Care was intended to strengthen the collective identity of the chemical industry by showing that the industry could control its environmental behavior voluntarily. In contrast, ISO 14001 is not a collective action program. The facilities that adopt the standard have no formal relationship with one another. As yet, there is no shared identity among firms that have adopted ISO 14001 or other standardized EMSs.

A recent analysis found that firms that participate in Responsible Care reduce their toxic releases no faster than comparable chemical firms that do not participate (King and Lenox 2000). The authors argue that the lack of mechanisms for observing and sanctioning individual firm performance has led to free riding by low-performing firms. Although some ACC members are improving environmental performance faster than the norm, a large group is lagging behind, slowing progress for the group overall.

Customer Demands. Whereas little evidence indicates that U.S. firms are under direct pressure from customers to adopt EMSs at this time, many people anticipate that such demands will strongly influence adoption in the future.[2] Firms that adopt EMSs in response to customer demands will establish systems that reflect customers' concerns. If customers want to ensure that their suppliers comply with environmental laws, then suppliers will adopt EMSs with compliance goals. If customers take the position that they will prefer companies that conserve natural resources and protect the interests of future stakeholders, then sustainability frameworks may be incorporated into EMSs.

Agency Negotiation. State and federal environmental agencies have begun to experiment with programs that offer regulatory benefits in exchange for the adoption of EMSs with specified characteristics. Such programs have created another set of motivations for firm EMS adoption. For example, agencies have required some firms to adopt standardized EMSs as part of settlement agreements. State performance track programs (such as those discussed in Chapter 9) and EPA's National Environmental Performance Track will offer regulatory flexibility to firms that meet certain criteria, including EMS adoption.

Why EMSs Fail to Ensure Excellence

A firm that adopts an EMS to strengthen its environmental performance probably will invest in environmental improvement. This investment is consistent with the factors that have driven its decision to adopt. However, firms motivated by other drivers will not necessarily make this investment choice.

The fact that EMSs are now being demanded by external constituencies such as customers, industry peers, and agencies has changed the relationship between the decision to adopt an EMS and subsequent decisions to invest in improvements. Managers face two separate decisions: whether to adopt an EMS and how highly to invest in environmental performance. Firms can adopt an EMS without investing in environmental performance improvement. This is the dilemma for policymakers who hope to use EMS adoption as a criterion for excellence.

Only certain kinds of EMSs require firms to invest in performance (for example, ISO 14001, with its continuous improvement basis). However, whereas ISO 14001 registration requires consistency between the formal EMS and practice, the goals that drive the system are left to the discretion of facility managers. The level of environmental investment will vary in accordance with the ambitiousness of the goals that managers choose.

In the next sections, we consider the organizational, strategic, and competitive context for EMSs. By *organizational context* we mean the way the system is embodied in the cultural structures of the organization (Ehrenfeld 1998). Managers and line workers interpret the formal EMSs within existing cultural meanings and values (Giddens 1984). Often there is a gap between the formal EMS and the informal rules, authority structures, and technical resources of the organization; this gap can have several causes. *Strategic context* refers to the business goals of the organization. The goal to achieve a certain rate of return can eclipse commitments to environmental performance improvement, especially during economic downturns. By *competitive context*, we mean the firm's relationship with suppliers, customers, and other firms in its industry. Environmental performance can become a basis for competition, but far more often, sales are made on the basis of cost and quality. Firms that stake out the ground of environmental leadership must regularly assess the degree to which environmental programs influence competitiveness.

Inconsistency between Official and Operative Goals

A conflict between the formal EMS and practice can occur when employees know that the management does not mean what it says. "Read my lips" is where they get the real message. The formal EMS exists for some other purpose, often as a signal to outsiders that managers have listened to their concerns.

EMSs often include a mix of official and operative goals. *Official goals* are those that are publicly stated by the organization and are often quite general (Switzer 1999). Their purpose is symbolic, and they provide the basis for the organization's ability to acquire legitimacy, allies, resources, and personnel (Milledge 1995). The guiding principles of Responsible Care represent official goals. *Operative goals* are specific enough to guide the behavior of organizational members: "They provide cognitive guidance and the basis for operating decisions at lower levels of the organization" (Milledge 1995, 11). Polaroid and Robbins set operative goals in the form of numeric reduction targets.

Failure of Goals to Embody Excellence

The goals established by facilities adopting ISO 14001 are by definition operative. The audit and corrective action procedures in ISO 14001 are designed to ensure that actual performance meets stated goals and objectives. Registrars frequently urge managers to establish goals that they are confident they will be able to meet.

A study of goals set by U.S. ISO 14001 facilities found a reduction in goal stringency after adopting this EMS standard (Switzer 1999). So, whereas ISO 14001 goals are operative, they may only require modest improvements in performance.

EMS Goals May Conflict with Strategic Priorities

An EMS may fail to lead to environmental excellence when normative parts of the EMS (that is, the values and priorities it espouses) conflict with other values and priorities in the organization. Actors may be forced to make trade-offs where environmental actions lose out (Ehrenfeld 1998).

When Polaroid's chemicals division determined that it could not meet its reduction target (10% per year) without what were considered excessively high costs, Polaroid changed its goals. On the other hand, Robbins realized cost savings from its EMS and increased its market share. These financial benefits allowed managers to continue to pursue their goals.

Slow Pace of Cultural Change

A final impediment to excellence is the slow pace of cultural change (Ehrenfeld 1998). It takes time for the organizational culture to catch up and to integrate new formalities, power relationships, and tools. There always will be a lag between the dissemination of an environmental policy statement, employee training, or other activities required under ISO 14001, for example, and the development of new routines that reflect the changes in rules and resources. When these new parts of the structure become embedded, they must share space with the preexisting culture.

In ISO 14001 systems, with periodic audit and corrective action processes, differences should diminish with time because the whole procedure is designed to make deviations between stated goals and results disappear. ISO registration is awarded only after such a cycle has already been completed.

The Policy Challenge

Agencies are concerned about two subclasses of excellence. The first is excellence in terms of regulatory compliance. Agencies charged with the enforcement of regulations and other codified rules cannot condone anything less than complete compliance at all times. This ideal is rarely if ever attainable. The fundamental argument for granting some form of agency largess to firms that adopt satisfactory EMSs is that these firms will move closer to the ideal than those following normal practices. Agency resources then can be shifted toward ensuring that other firms are meeting the norms.

The second category of excellence is differential environmental performance in some activity beyond the requirements of regulatory mandates. At least in theory, firms will exceed the norms of environmental performance if they are given incentives through regulatory flexibility that they assess as valuable enough to justify the investment in supranormal behavior. Awards and public recognition by an agency might be such incentives.

All excellence programs involve negotiation between the agency and firm and result in some form of contract (Meyer 1999). Through a process of negotiation, a firm and an agency agree that the firm will do x and the agency will do y, where x and y refer to some specified observable action. Project XL is an example of such a negotiated agreement. A few years ago, Digital Equipment Corporation entered into a Project XL agreement with U.S. EPA Region I. Digital agreed to produce a *Design for Environment* primer and related training materials; the agency promised to make a public announcement that Digital had been selected for the Project XL program.

Agencies must justify programs that exceed compliance on the grounds of their legal mandates. One argument for this form of excellence program is that companies willing to devote resources to beyond-compliance activities also will outperform the norm with respect to actions clearly within the agency's mandates. The agency benefits through more effective allocation of resources.

Excellence programs, based as they are on negotiated agreements, have the potential to break down over each party's difficulty in determining whether the other has kept its promise. Knowing how well another party has kept its promise is rooted in two characteristics: how well the conditions of satisfaction can be specified (*specificity*) and how well these conditions can be observed (*observability*). A third problem is how to handle breaches or failures to keep the promises (*enforceability*). The same problems of specificity, observability, and enforceability arise in standard contract theory in both legal and economic contexts.

Agency Response

The problem facing agencies corresponds to the typical principal–agent situation that has been intensely studied in the business economics field (Holmstrom and Milgrom 1994). How will a principal (the one making the request) know that the agent (the actor promising to fulfill some condition of satisfaction) is doing what was promised when the principal cannot practically observe the agent at work? Many solutions in business involve creating compensation schemes that provide adequate incentives (motivation) such that the agent will do what was promised without continuous supervision and observation.

The role of EMSs in excellence programs is a special principal–agent problem. The outside stakeholder—in this case, a government agency—is in essence requesting that a firm perform better than the norm for its type of business. The agency is the principal, and the firm is the agent. The government organization cannot easily observe such performance; monitoring is expensive, and the agency has a multitude of agents to monitor. The potential "solutions" follow from principal–agent theory. Government can pick some proxy for the performance they want, and offer a set of incentives to create the motivation needed.

An EMS is the proxy for environmental performance, and a package of valuable offerings by the agency is the incentive. The basic idea behind management standards such as ISO 9000 and ISO 14001 is that improved performance will follow from the implementation and operation of the EMS. Both of the ISO standards are grounded in continuous improvement models in which progress toward some explicit target is monitored periodically and changes are introduced when that progress is deemed to be inadequate. The circular continuous improvement process is frequently abbreviated as "plan, do, check, act."

Planning sets broad policies, specific targets, and strategic allocations of authority and resources. *Doing* puts the strategies and plans into play. *Checking* is stopping the process and examining how well the targets are being met on the basis of carefully collected information. If performance is on track, then the organization continues to repeat the doing and checking. If deviations are found, then *acting* involves going back and revising the plans so that satisfactory progress is made.

The effectiveness of EMSs in solving the observability problem in the principal–agent case depends on how well they predict performance. A firm's cultural, strategic, and competitive context strongly influences the degree to which the formal EMS will improve performance. Because of this uncertainty, agencies require explicit means of checking how the EMS is working in practice. Methods include direct monitoring by the principal, self-reporting by the agent, third-party auditing, or the use of public reporting under circumstances where interested stakeholders would convey concerns to the principal.

In response to uncertainties associated with each of the EMS types discussed earlier, to date, agencies have refrained from awarding "excellent" status to firms on the basis of EMS implementation alone. Instead, they have begun to develop programs to recognize and reward facilities that adopt EMSs with specified criteria. One agency program that offers insights for this discussion is EPA's StarTrack Program. StarTrack was initiated by U.S. EPA Region I in 1996 as a pilot project. The program's initial goal was to test using corporate compliance and EMS audits, verified by a third party, as a substitute for regular agency inspections. To be admitted into the program, a facility needed to have a strong compliance history, a track record of pollution prevention, and a formal EMS in place. The facility had to disclose its compliance and EMS audits to the public and publish an annual environmental performance report.

Participation in StarTrack never grew beyond a handful of firms. After two years, 15 facilities had joined. The perception on the part of firms was that EPA was not able to provide the benefits—public recognition, fast-track permitting, and inspection relief—that it had promised. Also, according to facility managers, the administrative costs associated with participation were significant.

By requiring facility managers to publicly disclose audit results, StarTrack offered a model for observing facility environmental performance: corporate and third-party audit reports were made available for review to anyone who asked; agency staff members usually observed audits, and in several cases, staff members of environmental advocacy organizations took part as well. EPA did less well in specifying its expectations for participants. EPA clearly expected regulatory compliance, but its expectation for beyond-compliance performance was unclear. As a result, agencies at times expressed disappointment in facility environmental performance, but managers were not always aware of how they were falling short. From the perspective of facility managers, the benefits of participation were not adequately clear. The benefits listed in StarTrack promotional materials went beyond those in Star-Track agreements signed by facility managers and agency staff. And, as alluded to earlier, facility managers were unable to enforce their understanding of StarTrack benefits. Facility managers had little leverage to use with agencies. Once admitted into the "excellent" tier, it was difficult to exit without losing stature. For these reasons, participation remained low.

In June 2000, EPA launched the National Environmental Performance Track to consolidate and build on StarTrack and other innovative programs such as the Environmental Leadership Program and Project XL. Performance Track will offer two levels: Achievement Track for facilities with strong environmental performance and Stewardship Track for facilities that achieve even higher levels. Notably, Performance Track does not require managers to have their EMSs audited by a third party, as did StarTrack. Instead, managers are allowed to conduct self-assessments of their EMSs and disclose results in an annual performance report. This change reduces the costs of participation for facilities. EPA's costs for administering the program, in terms of staff time required to review each facility's application, also are lower than for StarTrack. EPA will review Achievement Track applications only for completeness; it will not undertake a substantive review. EPA does not plan to inspect facilities to make sure they are following through on program commitments. It plans to visit the sites of approximately 20% of participating facilities.

The benefits that EPA will offer participating facilities are still being worked out. EPA has stated that benefits will involve no relaxation of substantive levels of compliance under current rules and must be relatively easy to implement on a national level. EPA will not offer benefits that have to be tailored to specific facilities. At this point, EPA is offering Achievement Track facilities recognition, enhanced access to decisionmakers in the agency, and reduced inspection frequency. Limited regulatory flexibility may be offered in the future. EPA recognizes that these benefits are modest. However, policymakers feel that the benefits are appropriate given the relatively low costs of participating in the program. The observability, specificity, and enforceability of Achievement Track agreements are not major EPA concerns, because it is relinquishing little in terms of its normal regulatory oversight of these facilities.

Conclusion

Although standardized EMSs are a relatively new policy tool, firms have used formal EMSs at least since the mid-1980s, when Polaroid and The Robbins Company implemented their systems. In understanding the role of EMSs in environmental policy, it is helpful to review these early experiences.

Firms that see environmental practices as marginal to their strategic and competitive objectives will treat EMSs as tools for external image manipulation and unimportant for internal change. Firms with strong environmental commitments will use EMSs as tools to become even stronger. Although little research has explored the role of EMSs in changing organizational culture, we suggest that EMSs serve primarily as reinforcing mechanisms. This finding holds true for practices introduced by firm-structured EMSs, trade association EMSs, and standardized EMSs.

The implications for policy of this discussion are fairly obvious. Two organizations that implement identical EMSs will realize very different results. Results depend on the motivation behind adoption and how the rules and resources demanded by the EMS are integrated into the culture of the organization.

Standardized EMSs require that firms adopt targets and objectives that are operative. Operative targets contribute toward specificity. However, operative goals do not necessarily embody excellence, that is, performance above a normal baseline. Standardized EMSs appear not to encourage "stretch goals" such as the goal of zero discharge set by The Robbins Company. In addition, although firms must make continuous improvements toward goals established in the standardized EMS, the pace of progress may be very slow, thus failing to meet the criterion of excellence.

Third-party oversight is one option for standardized EMSs that provides agencies with observability. It is not unique to this EMS type, however. Some trade association EMSs also use third-party oversight. Responsible Care uses industry peers as third parties, and the ISO 14001 system uses independent technicians specifically trained for this purpose. Independent technicians may offer greater objectivity, but industry peers may offer greater knowledge of manufacturing processes and attention to performance.

Agency attempts to structure EMS requirements are leading to a fourth category to add to our typology. However, early experience with agency-structured EMSs leads us to question their feasibility. The costs to firms of adopting EMS elements defined by agencies are substantially greater than for other EMSs, and so far, agencies have encountered difficulty keeping their end of the agreements. Participation will likely remain low until these issues can be addressed.

Notes

[1] One element that is emphasized in environmental management systems (EMSs) specifications today but was missing from the Polaroid and Robbins EMSs of the late 1980s is the practice of documenting procedures.

[2] During the study period, Ford and General Motors announced that they would require suppliers to adopt EMSs.

References

Berube, M., J. Nash, J. Maxwell, and J. Ehrenfeld. 1992. From Pollution Control to Zero Discharge: How The Robbins Company Overcame the Obstacles. *Pollution Prevention Review* Spring: 189–207.

CMA (Chemical Manufacturers Association). 1993. *On the Road to Success: Responsible Care Progress Report*. Arlington, VA: Chemical Manufacturers Association.

———. 1998. *Responsible Care Guiding Principles*. Arlington, VA: Chemical Manufacturers Association.

Ehrenfeld, J.R. 1998. Cultural Structure and the Challenge of Sustainability. In *Better Environmental Decisions*, edited by K. Sexton, A. Marcus, K.W. Easter, and T. Burkhardt. Washington, DC: Island Press.

Giddens, A. 1984. *The Constitution of Society*. Berkeley, CA: University of California Press.

Holmstrom, B., and P. Milgrom. 1994. The Firm as an Incentive System. *American Economic Review* 84: 972–991.

Howard, J., J. Nash, and J. Ehrenfeld. 2000. Standard or Smokescreen? Implementation of a Voluntary Environmental Code. *California Management Review* 42(2): 63–82.

King, A., and M. Lenox. 2000. Prospects for Industry Self-Regulation without Sanctions: A Study of Responsible Care in the Chemical Industry. *The Academy of Management Journal* 43(4): 698–716.

Meyer, G.E. 1999. *A Green Tier for Greater Environmental Protection*. Madison, WI: Wisconsin Department of Natural Resources.

Milledge, V. 1995. *Goal Setting and Task Performance at the Organizational Level: Studies of Emissions Reductions Goals and Performance.* Unpublished doctoral dissertation. Berkeley, CA: University of California, Berkeley.

Nash, J. 1997. ISO 14000: Evolution, Scope, and Limitations. In *Implementing ISO 14000*, edited by T. Tibor and I. Feldman. Chicago, IL: Irwin Professional Publishing, 495–510.

———. 1999. *The Emergence of Trade Associations as Agents of Environmental Improvement.* Draft report prepared for U.S. EPA Emerging Strategies Division. Cambridge, MA: Massachusetts Institute of Technology.

Nash, J., K. Nutt, J. Maxwell, and J. Ehrenfeld. 1992. Polaroid's Environmental Accounting and Reporting System: Benefits and Limitations of a TQEM Tool. *Total Quality Environmental Management* August: 3–15.

Nash, J., J. Ehrenfeld, J. MacDonagh-Dumler, and P. Thorens. 1999. *ISO 14001 and StarTrack: Assessing Their Role in Environmental Performance Improvement.* Washington, DC: National Academy of Public Administration.

Rees, J. 1997. Development of Communitarian Regulation in the Chemical Industry. *Law and Policy* 19(4): 477–528.

Simmons, P., and B. Wynne. 1993. Responsible Care: Trust, Credibility, and Environmental Management. In *Environmental Strategies for Industry: International Perspectives on Research Needs and Policy Implications*, edited by K. Fischer and J. Schot. Washington, DC: Island Press.

Switzer, J.K. 1999. *ISO 14000:* Regulatory Reform and Environmental Management Systems. Master's thesis in technology and policy. Cambridge, MA: Massachusetts Institute of Technology.

Uzumeri, M. 1997. ISO 9000 and Other Metastandards: Principles for Management Practice? *The Academy of Management Executive* 11(1): 21–36.

4

Why Do Firms Adopt Advanced Environmental Practices (And Do They Make a Difference)?

Richard Florida and Derek Davison

Since the dawn of the industrial revolution, the goals of economic growth and enhanced environmental quality have been at odds. The rise of industries such as steel, chemicals, automobiles, and electricity increased wealth, productivity, and profit but also brought about adverse environmental outcomes. This stark trade-off between economy and environment was particularly evident in industrial regions, which grew and prospered around resource extraction and heavy manufacturing. In such places, the environment often was seen as something that could be sacrificed in the pursuit of economic growth.

Out of this context grew the aggressive environmental policies and regulations of the late twentieth century, which addressed the economy–environment relationship by imposing strict limits on the waste and emissions produced by manufacturing companies (Andrews 1999). These approaches certainly produced important gains; however, as leading commentators around the world have noted, this command-and-control approach to environmental policy may have reached the point of diminishing returns (Strasser 1996).

Leading corporations in the United States and around the world are pioneering new strategies for integrating the environment into their overall business strategy and for simultaneously improving their environmental and business performance (Schmidheiny 1992; Hart and Ahuja 1994; Porter and van der Linde 1995a, 1995b). These firms are, to turn a phrase, becoming "leaner and greener" at the same time (Florida 1996; Florida and others forthcoming). These companies are motivated not by altruistic concerns but by the bottom-line drive to increase profits, productivity, and performance by reducing waste and emissions. Around the world, a "three-zero" manufacturing paradigm is emerging, where companies simultaneously work to achieve zero defects (quality), zero inventory (just-in-time inventory and supplier relations), and zero waste and emissions.

This trend toward increased business innovation has been well documented. However, little research exists regarding the ways that innovative plants interact with their surrounding communities. Are high-performance manufacturing facilities more likely to be good corporate citizens than their less innovative counterparts? Do these advanced factories use aspects of their internal innovation in their dealings with community leaders and residents? Are the principles of advanced manufacturing being applied beyond the factory walls?

In this chapter, we examine such issues within the context of environmental programs and performance. We explore whether environmentally advanced factories are more likely to maintain an open relationship with the surrounding community and whether an open relationship leads to a more positive, constructive association between the parties. It examines what, if anything, communities stand to gain from attracting innovative factories and encouraging existing facilities to innovate as well.

We examine a new and innovative approach to managing business goals and environmental performance: the rise of environmental management systems (EMSs). Like other management systems, EMSs are formal systems for articulating goals, making choices, gathering information, measuring progress, and improving performance. They are increasingly recognized as systematic and comprehensive mechanisms for improving environmental and business performance. In many ways, EMSs represent an extension of the core principles of total quality programs to environmental management. Government policymakers are interested in EMSs as a possible supplement to or replacement for so-called command-and-control environmental regulation. Additionally, we examine the benefits of pollution prevention programs used in conjunction with EMSs. The U.S. Environmental Protection Agency (EPA) defines *pollution prevention* (P2) as "source reduction—preventing or reducing waste where it originates, at the source—including practices that conserve natural resources by reducing or eliminating pollutants through increased efficiency in the use of raw materials, energy, water, and land."

Although research is available on environmental regulation and the relationship between environmental regulation and the innovation and performance of firms, very little serious empirical work has been done on EMSs specifically. The existing literature is based largely on case studies of environmental success stories, which provide valuable best-practice data but offer "virtually no evidence on the extent of the penetration of advanced practices across the industrial landscape and of the factors that influence their adoption and diffusion" (Florida 1996). Moreover, almost no empirical evidence exists about the community impacts of adopting EMSs and P2 programs. Research has concentrated on the effects of environmental improvement on internal factory performance with little emphasis on the ways that such improvement can positively or negatively affect the surrounding community and its relationship with the plant. We intend to try to close part of this gap by presenting empirical evidence regarding two key questions:

- To what degree are companies adopting EMSs (and P2) practices? What factors motivate companies to adopt such practices—regulations, business performance, commitment to advanced management, or concern for their environmental impacts on communities?
- What are the impacts of EMSs on plants and on communities? Do plants with advanced environmental practices (EMSs and P2) pose less environmental risk and confer more significant benefits on their communities than plants without such practices? To what degree are advanced environmental plants—that is, plants with EMS and P2 practices—better able to involve community stakeholders in these issues?

To answer these questions, we report the findings of new empirical data from our survey of more than 580 manufacturing plants (see Appendix for more detail). The survey, administered between September 1998 and February 1999, amassed information about

- plant characteristics (facility size, industry, and number of employees);
- adoption of advanced environmental and management practices (EMSs, P2, quality management, ISO 14000, and so forth);
- community environmental activities, modes of information sharing, and mechanisms for obtaining community input on setting environmental priorities and sharing information; and
- environmental impacts on the community (for example, waste and emission streams, noise, odor, and employment).

Where possible, surveys were completed by plant environment, health, and safety (EHS) personnel. In plants without EHS departments, plant managers were asked to complete the survey.

A wide range of manufacturing industries are represented in the sample. The top three industries represented are chemicals and allied products (12.6%), primary metals (10.3%), and fabricated metals (10.3%). Electrical and electronic machinery (7%); rubber and plastic products (7%); paper and allied products (6.5%); electric, gas, and sanitary services (6.5%); and nonelectrical machinery (6.1%) each made up more than 5% of the sample. Fourteen industries totaled 1% or more of survey respondents.

Field research was conducted to supplement and extend the findings of the survey research. Because of limited time and resources, it was impossible to develop a large number of field research sites across the entire distribution of plants or to use control groups or matched pairs of plants. Therefore, field research sites were selected from survey respondents that had adopted advanced environmental practices (for example, EMS and P2 programs). Sites also were selected to account for different sizes of plants in different kinds of communities. A team of two social scientists and an engineer with expertise in plant production and environmental and waste emissions control technologies conducted site visits. Interviews were conducted with plant management and environmental staff to obtain information about plant characteristics and environmental performance. Interviews also were conducted with community officials, community residents, government agency personnel, and local government leaders to obtain additional information about the impacts of plant practices on the community.

It is important to remember that the findings reflect mainly the perceptions of manufacturing plant managers on the nature of their community environmental performance. Although the field research did include interviews with community representatives, the survey was based on the viewpoints of only manufacturing plant managers. These perceptions may not accurately reflect community issues but should be viewed as a first step toward assessing that reality. Additionally, the nature of the survey sample (see Appendix) and the voluntary nature of the survey would suggest an upward bias in the research findings. The research findings should be interpreted with these caveats in mind.

The remainder of this chapter is organized as follows. In the next section, we report the findings regarding why firms adopt EMSs and other innovative practices. In the third section, we examine the impact of EMS practices on plant-level environmental performance. In the fourth section, we report the findings on the effect of EMSs on community "environmental citizenship" and discuss environmental impacts on the community. In the final section, we outline the implications of these findings for businesses, communities, and government.

Why Do Plants Adopt EMS and P2 Practices?

Around the country and the world, companies are moving to adopt advanced environmental practices that bolster both environmental performance and competitiveness. EMSs are increasingly recognized as systematic and comprehensive mechanisms for improving environmental and business performance. The survey collected detailed information about the kinds of practices plants are using and the reasons they are adopting them.

EMS and P2 Adoption

First, a relatively large number of factories are using EMSs.[1] We estimate that roughly a quarter (24%) of manufacturing plants with more than 50 employees have adopted an EMS, 28% have adopted a formal P2 program, and 18% can be classified as advanced or high-adopter plants (EMS/P2), that is, plants that have adopted both an EMS and a P2 program (Table 4-1).[2]

EMS/P2 and nonadopter plants are significantly different in terms of size and dedicated environmental resources (Table 4-1). For example, EMS/P2 plants are considerably larger than other plants, with an average of 250 more employees than other factories in the survey. EMS/P2 plants have more than four times the EHS staff of other plants and more than three times the dedicated environmental staff. Also, EMS/P2 plants tend to be owned by corporations, whereas other plants are more often independently owned.

These findings suggest that EMS/P2 plants have a greater pool of internal and corporate-level resources to devote to improving environmental performance. This suggestion is in line with the findings of previous research, which indicates that resources are an important and determinant factor in the adoption and effectiveness of advanced EMSs (see Florida 1996; Florida and others forthcoming).

What Motivates Plants to Adopt?

What motivates plants to adopt EMS and P2 programs? Do they adopt these programs to meet government standards, to improve their environmental

Table 4-1. Key Characteristics of Sample Plants

Characteristic	Total plants (n = 214)	EMS/P2 plants (n = 62)	Other plants (n = 99)
EMSs	42.1%	NA	NA
P2	40.7%	NA	NA
EMS & P2	29.0%	NA	NA
No. of employees	912.4	910.7	677.2
EHS staff[a]	6.0	8.1	2.0
Environmental staff[b]	4.2	4.5	1.4
Independently owned[a]	43.0%	27.4%	58.6%
Part of multiplant company[a]	47.7%	72.6%	31.3%

Notes: P2 = pollution prevention; *n* = number of plants; NA = not applicable.

[a] Significant at the 0.01 level.   [b] Significant at the 0.05 level.

performance, or to be more efficient and competitive? The data suggest that business benefits are an important motivation for adopting innovative environmental practices (Table 4-2). The three most commonly identified motivators according to survey results are "commitment to environmental improvement" (91.9%), corporate goals and objectives (88.7%), and business performance (87.1%). These motivators are followed by more traditional motivators such as improved community relations (85.5%), state regulatory climate (85.5%), and federal regulatory climate (83.9%).

EMS/P2 plants were asked to rank these factors from 1 to 6, where 1 is most important. They rank corporate goals (2.07), commitment to environmental improvement (2.54), state regulatory climate (3.13), business performance (3.46), federal regulatory climate (3.50), and improved community relations (4.38) as the most important factors in motivating them to adopt EMS and P2 programs.

In addition to collecting information about the factors that motivated plants to adopt EMSs, the survey asked plants about the factors that motivated their environmental initiatives in general in an attempt to compare plants that had an EMS with those that did not. More than 90% of EMS/P2 plants report factors such as regulatory compliance (100%), cost savings (100%), improved business performance (96.8%), and self-motivation (93.5%) as important. Other widely reported factors include employee concerns (88.7%), customer relations (88.7%), and community concerns (87.1%).

Asked to rank the factors that motivated them to improve their environmental performance, EMS/P2 plants report the following factors as most important: regulatory compliance (1.63), cost savings (3.82), self-motivation

Table 4-2. Factors Motivating EMS and P2 Adoption

Factors	Total (%) (n = 214)	EMS/P2 plants (%) (n = 62)	Other plants (%) (n = 99)
Commitment to environmental improvement[a]	56.1	91.9	22.2
Corporate goals and objectives[a]	55.1	88.7	20.2
State regulatory climate[a]	54.2	85.5	23.2
Federal regulatory climate[a]	53.7	83.9	24.2
Economic benefits or business performance[a]	52.3	87.1	19.2
Improved community relations[a]	51.9	85.5	21.2
Other	1.9	3.2	0.0

Note: n = number of plants.   [a] Significant at the 0.01 level.

(3.95), and improved business performance (4.68) followed by community concerns (5.37), employee concerns (5.67), and customer relations (5.67).[3]

Certainly, the data suggest that although business concerns are important motivators, regulatory compliance cannot be dismissed. Andrews and others (see Chapter 2) have found that regulatory compliance is still the single most important factor in motivating EMS adoption. Although the inclusion of nonmarket and government facilities in the Andrews sample may have driven regulatory numbers upward,[4] these findings are not incompatible with our findings. Regulatory compliance would seem to remain the single most important motivating factor, followed by a cluster of business-related concerns.

Are EMS Plants More Advanced in General?

Another important question to ask is whether EMS plants are more advanced than other plants—that is, do plants that adopt EMS and P2 practices also tend to adopt other advanced management practices, such as total quality management?

Some studies suggest that advanced environmental practices reflect a corporate commitment to advanced management in general (see Florida 1996; Florida and others forthcoming). Other studies indicate that advanced plants tend to adopt an interrelated bundle of advanced practices such as team-based work, employee input in decisionmaking, quality management, and so on (see Ichniowski and others 1997; MacDuffie 1994; Florida and Jenkins 1998; Jenkins and Florida 1999). To probe these issues, the survey looked at the adoption of a wide range of advanced management practices, such as ISO 9000 and 14000 certification, employee involvement in shop-floor decisionmaking, internal and external environmental audits, total quality management, and just-in-time inventory control. The survey findings (reported in Figure 4-1) indicate that EMS/P2 plants are significantly more likely to adopt a wide range of advanced or innovative practices.

According to the results of the survey, EMS/P2 factories are significantly more innovative in general. Nearly twice as many EMS adopters have total quality management programs and just-in-time systems. Adopters also are nearly twice as likely to make use of internal environmental audits than other plants. Additionally, EMS/P2 plants are far more likely (about 17 times as likely) to be ISO 14000–certified than are other plants.

Environmental Performance Measures

Environmental performance measures are mechanisms by which companies can jointly improve their environmental and business performance. Florida and others (forthcoming) find that environmental performance measure-

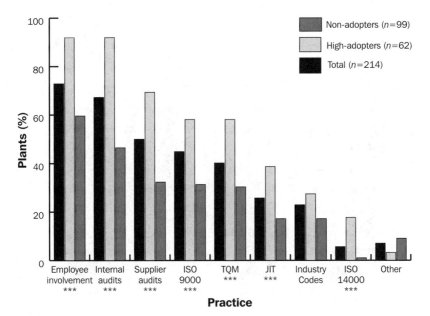

Figure 4-1. Adoption of Advanced Practices by Sample Plants

Notes: TQM = total quality management programs; JIT = just-in-time inventory.
***Significant at the 0.01 level.

Source: Florida and others forthcoming.

ment systems are key factors in effective implementation of environmentally conscious manufacturing systems. EMS/P2 adopters are significantly more likely to report using performance measures to track and monitor regulatory compliance, waste and emissions, and customer and community satisfaction (Figure 4-2).

Adoption of a Bundle of Advanced Practices

Summary scores were developed to look at the adoption of advanced environmental and management practices overall (Table 4-3). EMS/P2 plants are much more likely to implement the entire bundle of advanced environmental and management practices (both results are statistically significant at the 0.01 level).

■ EMS/P2 plants score nearly three times higher than other plants in the use of advanced management practices: 59.38 versus 21.40.

■ EMS/P2 plants are nearly twice as likely as other plants to make use of advanced management systems that consist of both advanced management practices and performance measurement systems: 64.85 versus 34.78.

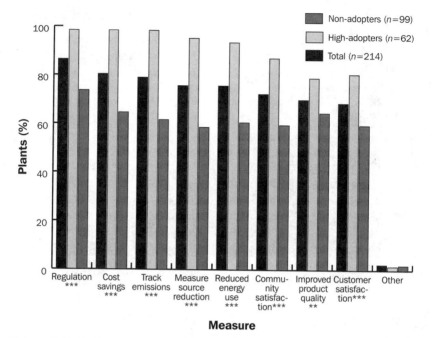

Figure 4-2. Environmental Performance Measures by Sample Plants

Notes: ***Significant at the 0.01 level; **significant at the 0.05 level.

Source: Florida and others forthcoming.

The summary scores thus suggest significant symmetry between the adoption of advanced environmental practices and the adoption of advanced management practices more generally. EMS/P2 plants are likely to adopt a broad bundle of innovative practices such as EMS, P2, total quality management, employee involvement, and performance measurement systems. Adoption of EMS and P2 practices is thus associated with plants that are larger and more innovative overall.

Plant Level Environmental Impacts

In addition to identifying the factors associated with why companies adopt EMSs, we examined the impact of EMSs on environmental outcomes. The survey asked companies to identify the ways in which EMSs affect their environmental performance. The results are clear: EMS plants are nearly twice as likely to report P2 as a source of plant-level improvement (93.5% versus 69.7%) and three times more likely to view EMSs as the source of significant in-plant improvement (79% versus 28.3%). They also are significantly more likely than other plants to cite the following as sources of plant level environmental improvement: recycling (93.5% versus 69.7%), air emission reduction (88.7% versus 53.5%), solid waste reduction (75.8% versus 54.5%), and electricity use (67.7% versus 43.4%).[5]

Are EMS Plants Better Environmental Citizens?

We wanted to know whether EMS plants were better environmental citizens than plants without EMS programs. That is, do EMS/ P2 plants tend to involve community groups in their environmental planning, sponsor environmental initiatives, or share information in ways that go beyond what other plants do? Studies have shown that advanced management systems are associated with interrelated systems of practices that foster information sharing, promote teamwork, and cultivate employee input in decisionmaking (Ichniowski and others 1997; Jenkins and Florida 1999). We wanted to see to what degree advanced plants reflected these underlying practices in their dealings with communities. To what extent did plants with advanced environmental practices share information with communities or involve community groups in the design and development of relevant environmental initiatives?

Information Sharing

The survey findings indicate that EMS plants are more likely to share information about their environmental practices with various groups (government agencies, businesses, neighbors, community groups, environmental groups, and so on) (Figure 4-3) than are plants without EMS programs. EMS/P2 plants are significantly more likely to report sharing information with government agencies, business customers, neighbors, and environmental groups. They are more than twice as likely to share information with neighbors and almost four times more likely to share information with environmental groups. EMS/P2 plants are four times more likely to engage in private meetings with community leaders and more than six times more likely to engage in broad-based community meetings. They also are much more likely to make use of Citizens' Advisory Councils or to survey local citizens.[6]

Community Involvement

Another way that plants can extend the basic principles of advanced environmental practices to communities is by involving community groups in environmental programs. The survey examined whether EMS/P2 plants are more

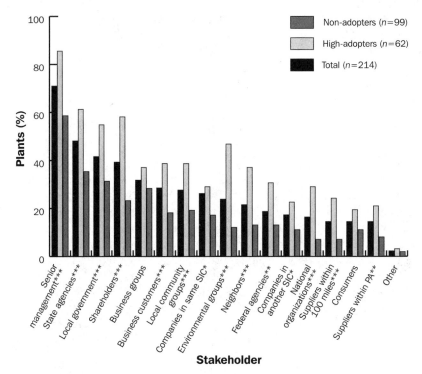

Figure 4-3. Environmental Information Sharing by Sample Plants

Notes: SIC = Standard Industrial Classification; PA = Pennsylvania. ***Significant at the 0.01 level; **significant at the 0.05 level; *significant at the 0.1 level.

Source: Florida and others forthcoming.

likely than other plants to involve communities in environmental activities and setting priorities (Figure 4-4). EMS plants are almost three times as likely to involve neighbors and citizens and more than twice as likely to involve local government. They also are statistically significantly more likely to involve community groups, environmental groups, and local businesses.

Sponsoring Environmental Activities

EMS/P2 plants also are more likely to engage in a wide range of environmental activities with their local communities than other plants are. They are three times more likely to sponsor educational awareness programs and two times more likely to sponsor recycling programs or Earth Day events. They are four times more likely to sponsor neighborhood beautification programs, and more than 10 times more likely to provide grants for local environmen-

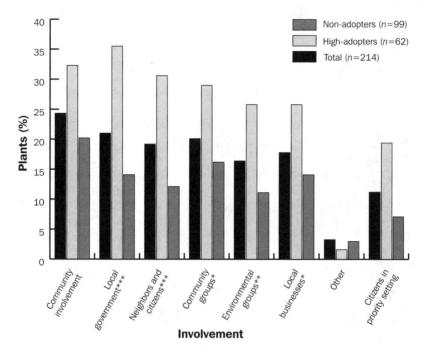

Figure 4-4. Community Involvement in Environmental Program Development by Sample Plants

Notes: ***Significant at the 0.01 level; **significant at the 0.05 level; *significant at the 0.1 level.

Source: Florida and others forthcoming.

tal projects and activities. Indeed, EMS/P2 plants devoted considerably more financial resources to community environmental activities than other factories, an average of $12,750 per plant versus $5,666 per plant.

The survey also asked for respondents' perceptions of the change in their relationship with the surrounding community over the preceding five years; respondents were asked if they would classify relations as "much improved," "improved," "unchanged," "worse," or "much worse." Nearly three-quarters of EMS/P2 plants (72.4%) reported that their relationship with the surrounding community was either "improved" or "much improved," whereas only 44.6% of other plants gave those answers (Figure 4-5).

Another way to gauge the nature of the relationships between manufacturing plants and their surrounding communities is to examine the way that communities react to potentially sensitive proposals from those plants. The survey sought to obtain plant managers' perceptions of how the community has responded to major plant initiatives such as permit applications, permit

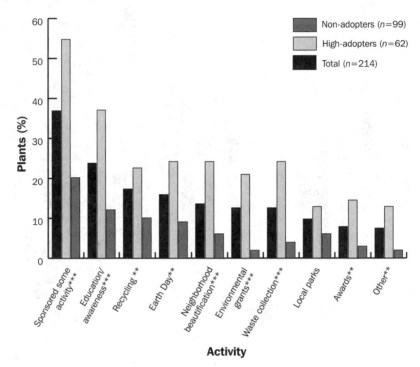

Figure 4-5. Community Environmental Activities by Sample Plants

Notes: ***Significant at the 0.01 level; **significant at the 0.05 level.

Source: Florida and others forthcoming.

revisions, and plant expansions.[7] EMS/P2 plants are much more likely to characterize their relationships with communities as "supportive" across each of these dimensions. Between 80% and almost 90% of EMS/P2 plants report that relationships with their communities were supportive in these cases, compared with 60–66% of other plants.

Community Involvement: Overall Trend

Summary scores were developed to look at the overall picture of community involvement and relationships (Table 4-3). The findings indicate that EMS/P2 plants consistently outperformed other plants on this dimension (all these results are statistically significant).

- EMS/P2 plants score three times higher than other plants in sponsoring community activities (23.55 versus 7.27), twice as high in sharing information with the community (37.48 versus 18.60), and more than twice

Table 4-3. Summary Scores for Sample Plants

	EMS/P2 plants (n = 62)	Other plants (n = 99)
Advanced practices		
Assessment[a]	70.31	48.17
Management[a]	59.38	21.40
Overall[a]	64.85	34.78
Community relations		
Activities[a]	23.55	7.27
Involvement[a]	16.56	6.77
Information sharing[a]	37.48	18.60
Overall[a]	32.65	17.31
Community environmental impact		
Direct impact[b]	28.19	20.71
Indirect impact[b]	32.54	24.04
Overall[a]	30.36	22.37

Note: n = number of plants.

[a] Significant at the 0.01 level.   [b] Significant at the 0.05 level.

- as high in involving the community in their environmental programs (16.56 versus 6.77).
- Overall, EMS/P2 plants outscored other plants by a factor greater than two: 32.65 to 17.31.

Generally speaking, EMS/P2 plants are more likely to share information with the community and to obtain input from community groups, neighbors, and environmental groups in making environmental decisions and setting environmental priorities. These community practices reflect the same basic principles of information sharing and employee involvement that underpin advanced management systems. We thus find that advanced plants are able to learn from and extend the principles of advanced management practices to their dealings with local communities.

Do EMS Plants Pose Less Environmental Risk for Communities?

The survey asked several questions to determine whether EMS/P2 plants are likely to pose less environmental risk and indeed to confer greater environmental benefits to the quality of the local environment than other plants. The short answer here again is yes. The survey data indicate that EMS/P2 plants consistently report posing less environmental risk and conferring more significant benefits in a wider range of areas than other plants.

Community Environmental Impacts

The survey asked a variety of questions designed to examine the environmental impacts of plants on communities. It focused on two types of environmental impacts: *direct environmental impacts,* which include waste emissions and energy use, and *indirect impacts,* or environmental esthetics and quality-of-life issues, such as reduced odor and improved plant appearance.

The survey results indicate that EMS/P2 plants pose less environmental risk and confer greater environmental benefits to their communities than other plants (Figure 4-6). First, EMS/P2 plants are much more likely to identify direct reductions in community environmental risk through a range of strategies for reducing and eliminating emissions and waste, which include programs for recycling, as well as controlling air emissions, solid waste disposal, energy use, fossil fuel use, and water pollution.

Second, EMS/P2 plants are more likely to report a positive impact on the local environment through improvements in plant and community esthetics such as reduced odor and dust or improved plant appearance. The importance of these kinds of improvements should not be minimized; the field research confirmed how important esthetic improvements can be to communities and residents. Nearly 60% of EMS/P2 plants report a positive impact through reduced dust compared with 39.4% of other plants. Two-thirds of EMS/P2 plants report a positive environmental impact through improved plant appearance compared with 55% of other plants. EMS/P2 plants are also more likely to report community benefits through increased local property values, 43.5% compared with 31.3% for other plants.

Third, one obvious way that any manufacturing plant affects the surrounding communities is by providing a source of employment for community residents. The survey asked respondents to estimate the overall effect of their environmental initiatives on plant employment, according to one of four outcomes: "eliminated jobs," "had no impact," "retained jobs that would otherwise have been eliminated," or "added new jobs." More than half of EMS/P2 plants (56.2%) report that they have added or retained jobs as a result of major environmental programs and initiatives compared with 26.3% of other plants.

Fourth, the findings suggest considerable symmetry and overlap between practices that result in both plant and community gains; this provides support for the hypothesis that advanced plants take what they have learned inside their factories and use it to realize improvements in community environmental impacts. Several of the most commonly identified sources of plant environmental impacts also are cited as sources of community envi-

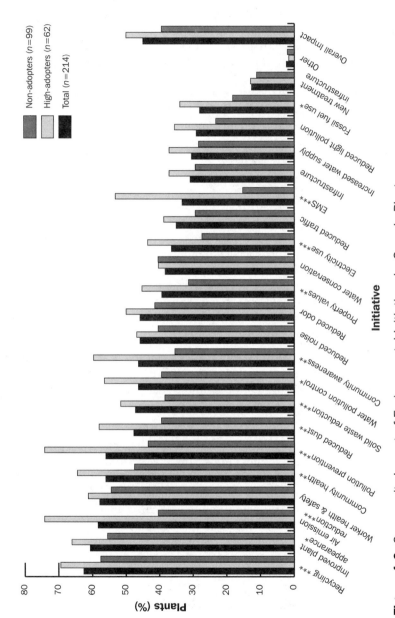

Figure 4-6. Community Impact of Environmental Initiatives by Sample Plants

Notes: ***Significant at the 0.01 level; **significant at the 0.05 level; *significant at the 0.1 level.

Source: Florida and others forthcoming.

ronmental benefit, especially recycling; P2; reduced air emissions, solid waste, and electricity use; and EMS.[8]

Fifth, EMS/P2 plants are much more likely to view having an EMS as a key factor in reducing community environmental risk. Three-quarters of these plants cite their P2 efforts as having a significant impact on community environmental quality compared with 43.4% of other plants. More than half of the EMS plants report their EMS as having a positive impact on community environmental quality compared with 15% of other plants.[9] Nearly 60% of EMS factories report their efforts to increase community awareness of P2 as having a significant positive impact on the environment compared with 35.4% of other plants. EMS/P2 factories are much more likely to rate their EMS as a source of community environmental improvement (3.43 versus 2.64).[10]

Motivating Factors

The survey obtained information about the key factors that plant managers believe influence their ability to have a positive effect on the community. It asked respondents to identify and rate the importance of factors such as environmental regulations; business leadership; advanced environmental programs, such as EMS and P2 programs; open sharing of information; constructive partnerships among business, government, and citizens; quality of local government; active citizen involvement; and active environmental groups.

The findings indicate that EMS/P2 plants are more likely than other plants to identify all these factors as important.

- More than three-quarters of EMS/P2 plants rate each of these factors as important compared with one-half to two-thirds of other plants.
- EMS/P2 plants in particular are much more likely than other plants to identify active citizen involvement and active environmental groups as important.

The survey also asked respondents to rate the importance of these factors on a scale of 1 to 4, where 4 is most important. The results here are somewhat similar across the EMS/P2 and nonadopter groups, not surprising given that these ratings seek to capture differences in the level of importance among plants that have already identified these factors as important. It is worth pointing out that the largest level of difference is found in the relative rating of "corporate EMS/P2 programs." Indeed, this is the only factor where there is a statistically significant difference between EMS/P2 and other plants. EMS/P2 plants rate corporate EMS/P2 programs a 2.98 compared with 2.35 for other plants. This score suggests that EMS/P2 programs represent a very important motivator for improving the local environment on the part of EMS/P2 plants. These findings lend additional confirmation to the hypothesis that EMS/P2 plants are extending the innovative practices and initiatives they use inside the plants into their dealings with communities.

Community Environmental Risk: Overall Trend

Summary scores were used to provide a broad assessment of the environmental risks and benefits plants pose for communities and to highlight the differential risks and benefits posed by EMS/P2 and nonadopter plants (Table 4-3).

■ The community impact scores indicate that EMS/P2 plants pose less environmental risk and confer potentially greater environmental benefits.
■ EMS/P2 plants outscored other plants in overall community impact (30.36–22.37), direct community impacts (for example, waste and emission reduction [28.19–20.71]), and indirect impacts (for example, environmental aesthetics [32.54–24.04]). All these findings are statistically significant.

Several of the most commonly identified sources of plant environmental impacts also are cited as sources of community environmental impacts: recycling, P2, air emissions reduction, solid waste reduction, electricity use, and EMS. There thus appears to be considerable overlap or symmetry in the practices that are sources of both environmental performance improvement inside the plant and of reduced environmental risk to communities.

Conclusion

Generally speaking, the survey findings suggest that EMS adopters are likely to be better corporate citizens and more innovative in their manufacturing processes than other plants. EMS facilities are more likely to share information with the community and to obtain input from stakeholders in making environmental decisions and setting environmental priorities. There is considerable overlap between EMS adoption and the adoption of other advanced management practices, such as total quality management and internal audits. The adoption of EMS and P2 practices seems to be associated with plants that are large and owned by corporations, suggesting that resource availability may play a key role in EMS adoption.

In short, the findings suggest that EMS/P2 adopters are extending the basic lessons and efficiencies generated by practices and initiatives originally implemented inside the plant to their relationships with and environmental impacts on communities. There appears to be a strong correlation between EMS adoption and the generation of positive environmental and economic outcomes for the community, but further research must be done to determine whether a causal relationship exists. The conclusion would seem to be that communities and governments have an incentive to encourage manufacturing facilities to adopt EMS and P2 programs. Perhaps addressing the resource discrepancy between EMS adopters and nonadopters would be one area for policymakers to explore.

Appendix: Research Design

This study is principally based on a survey of manufacturing establishments. The survey collected information regarding

- plant characteristics (plant size, industry, and number of employees);
- adoption of advanced environmental and management practices (for example, EMSs, P2, quality management, and ISO 14000);
- community environmental activities, modes of information sharing, and mechanisms for obtaining community input on environmental priority setting and information sharing; and
- community environmental impacts (for example, waste and emission streams, noise, odor, and employment).

The survey was administered between September 1998 and February 1999. The survey instrument was pretested with a small sample of manufacturing plants. Experts in survey design and community environmental impacts from academia, the consulting community, industry, environmental groups, and government agencies also reviewed the survey instrument.

Survey Sample

The survey was administered to a total of 583 manufacturing plants in Pennsylvania. The sample was designed to compare the environmental performance of advanced environmental plants with that of nonadvanced plants. To accomplish this goal, the overall sample was made up of three subsamples. The first subsample (n = 242 plants) was a random sample of all manufacturing plants in Pennsylvania, stratified by industry and size and selected from the *1998 Harris Directory* of manufacturing plants in Pennsylvania. Two additional subsamples were used to ensure that the sample included a significant number of plants that had adopted advanced environmental practices. A second subsample (n = 66 plants) was drawn from manufacturing plants that were recipients of the Pennsylvania Governor's Award for Environmental Excellence for the years 1996 and 1997. The third subsample (n = 275 plants) was made up of plants that had shown some interest in advanced environmental practices and was drawn from lists of manufacturing firms that had participated in regional Pollution Prevention Roundtables.

Survey Administration

The survey was administered by facsimile and then was followed up to maximize the response rate. Approximately two weeks after the initial fax, a second fax was sent to companies that had not yet responded. After another two to three weeks, follow-up phone calls were made to plants that had not yet responded. At that point, plants were given the option of being removed from the initial survey and being classified as nonrespondents. Of the 583 sample plants, 158 indicated that they were unwilling to participate in the survey. Of the remaining 425 plants, 214 responded to the survey for an adjusted response rate of 50.4%.

A wide range of manufacturing industries is represented in the sample. The top three industries represented are chemicals and allied products (12.6%), primary metals (10.3%), and fabricated metals (10.3%). Electrical and electronic machinery (7%); rubber and plastic products (7%); paper and allied products (6.5%); electric, gas, and sanitary services (6.5%), and nonelectrical machinery (6.1%) each made up more than 5% of the sample. Fourteen industries made up 1% or more of survey respondents. It is important to remember that the findings mainly reflect the perceptions of manufacturing plant managers on the nature of their community environmental performance. Although the field research included interviews with community representatives, the survey research was based on manufacturing plants only.

Field Research

Field research supplemented and extended the findings of the survey research. Because of limited time and resources, it was impossible to develop many field research sites across the entire distribution of plants or to use control groups or matched pairs of plants. Field research sites were selected from survey respondents that had adopted advanced environmental practices (for example, EMS and P2 programs). Sites also were selected to account for different sizes of plants in different kinds of communities.

Company reports and government documents were reviewed for plant background. Preliminary phone interviews were conducted with plant managers and environmental representatives to obtain information about environmental initiatives and community impacts and to ensure that the plants were viable field research candidates. Site visits of approximately half a day (including a plant tour) were conducted at each facility. A team of two social scientists and an engineer with expertise in plant production and environmental and waste emissions control technologies conducted these site visits.

Interview questions that covered plant practices, corporate practices, environmental performance, community relations, community impacts, and the factors associated with these initiatives were developed for each facility based on a review of reports, documents, and their completed survey form. Interviews were conducted with plant managers and environmental staff to obtain information about plant characteristics and environmental performance. Interviews also were conducted with community officials, community residents, government agency personnel, and local government leaders to obtain additional information about the impacts of plant practices on the community. More than two dozen interviews were conducted at the five sites.

Acknowledgements

Matthew Cline and Sam Youl Lee assisted with data collection and analysis and in the preparation of the study and this chapter.

Funding for data collection was provided by the Pennsylvania Department of Environmental Protection, Office of Pollution Prevention and Compliance Assistance, and the National Science Foundation Division of Geography and Regional Science and Environmentally Conscious Manufacturing Program, Award 9528766.

Notes

[1]Because the sample was composed of plants with environmental management systems (EMSs) in place and a control group, simply reporting the figures on EMS adoption for the entire sample would considerably overrepresent the extent of EMS adoption. The best estimate of EMS adoption in general comes from the findings for the control group that is a stratified random sample of plants with more than 50 employees.

[2]As Table 4-1 shows, the figures for adopting EMSs and pollution prevention (P2) programs for the survey sample are considerably higher than for the control group (which is a better representation of the population of plants). Almost 30% (29.0%) of plants in the entire sample are classified as "EMS/P2 plants"; that is, they had adopted both an EMS and a P2 program. Additionally, more than 40% of plants in the entire sample had either a formalized EMS (42.1%) or an active P2 program (40.7%). More than 45% (46.3%) of plants in the entire sample are classified as "other plants," meaning that they did not use either an EMS or a P2 program.

[3]Factors were ranked in order of importance, with one as most important.

[4]The research by Andrews and others shows that nonmarket facilities are more likely to identify regulatory compliance as a key motivator than are market facilities (see Chapter 2).

[5]Overall, plants with EMSs cited the following programs as major sources of in-plant environmental improvement: pollution prevention (95.2%), recycling (93.5%), reduction or elimination of air emissions (89.7%), worker health and safety (85.5%), EMSs (79.0%), and solid waste reduction and elimination (75.8%).

[6]Plants with EMSs were significantly more likely to share information via newsletters (43.5% versus 21.2%), local school programs (33.9% versus 12.1%), community relations departments (32.3% versus 7.1%), private meetings with community leaders (32.3% versus 8.1%), community meetings (32.3% versus 5.1%), the Internet (22.6% versus 5.1%), and citizen surveys (16.1% versus 2.0%).

[7]The survey did this to identify specific kinds of plant actions that could provoke a negative community reaction. It allowed respondents to identify recent sources of friction with their communities and served as a check against excessively positive estimates of community relations.

[8]As noted earlier, more than three-quarters of plants with EMSs and pollution prevention (P2) programs in place (EMS/P2 plants) cited pollution prevention, recycling, air emissions, worker health and safety, EMSs, and solid waste reduction and elimination as key sources of environmental improvement inside the plant. EMS/P2 plants were nearly twice as likely to report P2 as a source of plant-level improvement and were more than three times more likely to view an EMS as the source of significant in-plant improvement.

[9]Both of these results are statistically significant at the 0.01 level.

[10]On a scale of 1 to 4 (where 4 is most important and 1 is least important). This was the only response category that was statistically significant at the 0.01 level.

References

Andrews, Richard N.L. 1999. *Managing the Environment, Managing Ourselves*. New Haven, CT: Yale University Press.

Florida, Richard. 1996. Lean and Green: The Move to Environmentally Conscious Manufacturing. *California Management Review* 39(1): 80–105.

Florida, Richard, and Davis Jenkins. 1998. The Japanese Transplants in North America: Production, Organization, Location, and R&D. In *Between Imitation and Innovation: The Transfer and Hybridization of Production Systems in the International Automobile Industry*, edited by Steven Tolliday. Oxford, U.K.: Oxford University Press: 189–215.

Florida, Richard, Mark Atlas, and Matt Cline. Forthcoming. What Makes Companies Green? Organizational Capabilities and the Adoption of Environmental Innovations. *Economic Geography*.

Hart, Stuart, and Gautum Ahuja. 1994. Does It Pay to be Green? An Empirical Examination of the Relationship between Pollution Prevention and Firm Performance. Working paper. University of Michigan School of Business Administration.

Ichniowski, Casey, Kathryn Shaw, and Giovanna Prennushi. 1997. The Effects of Human Resource Management Practices on Productivity. *American Economic Review*.

Jenkins, Davis, and Richard Florida. 1999. Work System Innovation among Japanese Transplants in the United States. In *Remade in America: Japanese Transplants and the Diffusion of Japanese Production Systems*, edited by Paul Adler, Mark Fruin, and Jeffrey Liker. New York: Oxford University Press.

MacDuffie, John Paul. 1994. *Human Resource Bundles and Manufacturing Performance: Flexible Production Systems in the World Auto Industry*. Philadelphia, PA: Wharton School, University of Pennsylvania.

Porter, Michael, and Claas van der Linde. 1995a. Green and Competitive: Ending the Stalemate. *Harvard Business Review* September–October: 120–134.

———. 1995b. Toward a New Conception of the Environment-Competitiveness Relationship. *Journal of Economic Perspectives* 9(1): 97–118.

Schmidheiny, Stephen. 1992. *Changing Course: A Global Business Perspective on Development and the Environment.* Cambridge, MA: MIT Press.

Strasser, Kurt A. 1996. Preventing Pollution. *Fordham Environmental Law Journal.* 8(Fall): 1–57.

5

Environmental Management Systems and the Global Economy

Theodore Panayotou

An environmental management system (EMS), when properly implemented, can improve the condition of the environment and enhance the environmental performance of the organization that implements it. An EMS also can generate better relationships between the firm and its customers and other stakeholders, improved conditions for the local community, and streamlined and enhanced relationships with government agencies involved with the firm. In addition to these attributes, EMSs hold promise for producing global benefits that range from enhancing the gains from trade to a more active participation of developing country governments and firms in protecting the global environment.

In this chapter, I examine

- how an EMS can deliver global value and how it can help business simplify management of multiple jurisdictions;
- how an EMS relates to the developing world and how it transcends nation-states by using private-sector devices such as supply chains and insurance pools; and
- which attributes of an EMS deliver more global value, which are less appealing to shareholders, and which merit further development and propagation.

I conclude with research questions and data requirements for further development and deployment of an EMS on a global scale.

Does an EMS Deliver Global Value?

Global economic value may be defined as improved economic, environmental, and social performance (or conditions) beyond the individual firm, industry, or country. As such, global value may be understood as either the international implications of EMS adoption (especially as they relate to international trade and competitiveness) or the potential value of EMSs in managing "global commons" such as climate change and biodiversity. In this chapter, my focus is the former, but I also touch on the potential for the latter, especially in terms of emerging research questions.

In a world that is increasingly globalized and more environmentally aware, EMS implementation has the potential to deliver greater global value than (a) the total lack of international environmental standards that is the status quo or (b) uniform environmental standards applied worldwide, regardless of local conditions and stage of development of environmental infrastructure and instruments to encourage environmental protection beyond mere regulatory levels.

In this section, I examine the different ways in which an EMS could provide global value compared with the two alternatives mentioned earlier. Developed countries with relatively strict environmental standards favor the establishment of uniformly high international standards to level the playing field and to ensure that developing countries with relatively lax environmental standards or enforcement do not reap a competitive advantage in international trade. Although little evidence indicates that developing countries are becoming pollution havens for investment or trade (Panayotou and Vincent 1997; see Table 5-1 and Figure 5-1), the issue was very much on the agenda of the World Trade Organization (WTO) trade negotiations. Developing countries contend that developed countries are using environmental standards as nontariff barriers to trade and decry any efforts to impose developed country environmental standards on them, which they can ill afford. WTO rules, on the other hand, permit importing countries to reject products that do not meet their phytosanitary standards but do not allow them to impose process standards on the exporting countries. Against this background, implementation of EMSs offers opportunities to generate global value in both economic and environmental terms.

EMSs in general, and EMSs based on ISO 14001 or its sibling standards and guidance documents in particular, provide the framework for a voluntary but common and systematic set of procedures for improving environmental performance without dictating specific performance standards (IISD 1996). An EMS provides for setting priorities and targets, assigning responsibility for accomplishing them, measuring and reporting results, and some limited mechanisms for verifying claims externally. Although it calls for continuous improvement of the management system, it does not call for or

Table 5-1. Surveys of the Importance of Environmental Regulations to Plant Location in the United States

Survey	Sample	Result
Epping 1986	Survey of manufacturers (late 1970s) that located facilities 1958–1977	"Favorable pollution laws" ranked 43rd to 47th of 54 location factors presented.
Fortune 1977	*Fortune*'s 1977 survey of 1,000 largest U.S. corporations	11% ranked state or local environmental regulations among top five factors.
Schmenner 1982	Sample of Dun & Bradstreet data for new *Fortune* 500 branch plants opening 1972–1978	Environmental concerns are not among the top six items mentioned.
Wintner 1982	Conference Board survey of 68 urban manufacturing firms	29 (43%) mentioned environmental and pollution control regulations as a factor in location choice.
Stafford 1985	Interviews and questionnaire responses of 162 branch plants built in the late 1970s and early 1980s	"Environmental regulations are not a major factor" but more important than in 1970. When only self-described "less clean" plants were examined, environmental regulations were of "mid-level importance."
Alexander Grant (various years)	Surveys of industry associations	Environmental compliance costs are given an average weight of <4% but grow slightly over time.
Lyne 1990	*Site Selection* magazine's 1990 survey of corporate real estate executives	Asked to pick 3 of 12 factors that affect location choice, 42% of executives selected "state clean air legislation."

necessarily lead to improvements in environmental performance and does not set performance values (IISD 1996) or compliance schedules. Improvements can be slow or fast, toward high or modest goals (Welford 1996). It is a way of "systematically setting and managing performance commitments ... establishing 'how to' achieve a goal, not 'what' the goal should be" (IISD 1996).

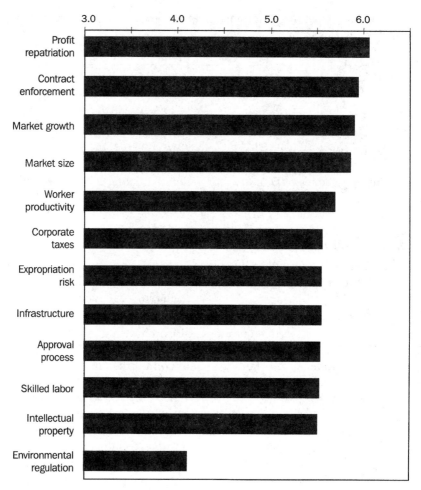

Figure 5-1. Comparative Impact on Industrial Location Decisions

Source: Panayotou and Vincent 1997.

As such, EMSs suit a world that is vastly different in terms of economic, social, and environmental conditions yet also experiences the common forces of globalization and increased environmental consciousness. Recognizing both commonality and divergence among countries, the global community has established "common but differentiated" responsibilities for developed and developing countries in the negotiation of international environmental treaties, such as the Montreal Protocol and the Framework Convention on Climate Change. A well-designed and earnestly implemented EMS is fully consistent with this concept (U.S. EPA 1998) and allows each country—indeed, each company—to choose its goals and pace of environmental improvement toward common goals (reducing pressure on the local and global environment and advancing the progress of sustainable development). "The National Technology Transfer and Advancement Act of 1995 (NTTAA) requires federal agencies to use voluntary consensus standards in certain activities as a means of carrying out policy objectives or other activities, unless the use of these standards would be inconsistent with applicable law ..." (U.S. EPA 1998).

The potential global values of worldwide implementation of a standardized EMS framework (such as that presented by ISO 14001 and its successor document, the forthcoming ISO 14001: 2000 revisions to the 1996 standard) by individual manufacturing, regulatory, and other organizations consist of

- gains from trade through the avoidance of nontariff trade barriers,[1]
- enhanced competitiveness of industry in developed and developing countries through more efficient production and induced innovation (see Figure 5-2),
- development benefits and continued economic growth in developing countries,
- gradual but continuous environmental improvement in developed and developing countries as well as increased contribution of all states to the protection of the global commons, and
- gradual convergence of economic and environmental conditions among developing and developed countries toward higher standards ("race to the top" rather than "race to the bottom").

Use of a standardized EMS framework by firms and other organizations may even facilitate the introduction of other common international environmental standards—environmental labels and declarations, life cycle assessment, and so forth. As long as companies in developing countries adopt a system of setting and managing performance commitments with shareholder involvement and external verification of claims, continuous environmental improvement, however slow, will produce a convergence in the long run. Most critically, extensive adoption of EMSs by firms, industries and organizations throughout the world has the potential to be the first concrete step of the world toward sustainability. However, EMSs may not themselves provide the motivation to firms to make costly changes; rather, public policymakers need to think about how to influence managers' motivations. A second critical area is that of international trade and the quest for a "level playing field" with regard to environmental standards, a subject that was part of the agenda of the Third WTO Ministerial Conference, held in Seattle, Washington, from November 30 to December 1, 1999. Yet another area of potential global value of EMSs is the implementation of international environmental agreements ranging from the Kyoto Protocol to the Convention on Biological Diversity.

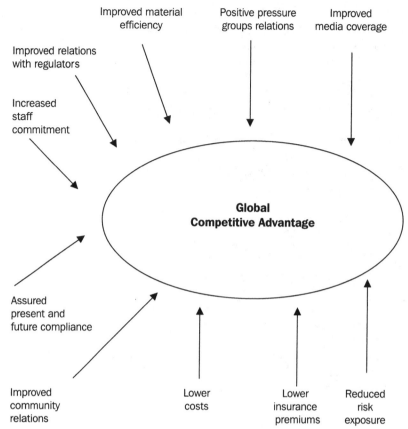

Figure 5-2. Potential Contributions of EMSs to Global Competitive Advantage

Source: Welford and Gouldson 1993.

How Does an EMS Help Businesses That Operate in Diverse Jurisdictions Simplify Their Management of Environmental Issues?

Globalization and the huge movements of private capital across borders—especially in the form of foreign direct investment—have multiplied the number of businesses that operate under multiple government jurisdictions beyond the limited number of multinational businesses that traditionally operated in more than one jurisdiction. Multinational businesses in different jurisdictions operate under different laws and regulations. Meeting the standards of different countries can be expensive. Local communities are sensitive to multinational businesses using different standards in different countries. At the same time, the global reputations of these companies depend on their environmental performance in each and every location.

Use of a framework EMS across the organization holds the potential to build consensus throughout global operations and provide a basis for systematically managing environmental risks in diverse geographies and in all locations. Kevin Wilson, CEO of Carter Holt Harvey International Corporation (New Zealand), said, "[EMSs] may result in considerable savings by reducing the cost of obtaining permits around the world and enabling the company to obtain longer term permits" (Wilson 1996). Such a common terminology improves the efficiency of communications. If a common international EMS framework such as that offered by ISO 14001 is adopted by an organization in all the jurisdictions in which it operates, it might simplify environmental management and reduce transaction costs (Welford 1996). A unified approach to environmental management provides opportunities for sharing ideas and practices among facilities across borders, increasing efficiency, and reducing costs.

One major source of tension between multinational and local firms is the different environmental standards to which they must adhere. Multinational businesses, for the reasons given earlier, often are compelled to comply with more country standards, whereas local competitors fall under the less stringent standards of the host country. Unlike command-and-control regulations, which must apply equally to all firms within a jurisdiction, EMSs permit differentiation between multinational and local firms. An EMS helps provide the uniformity (in process) across jurisdictions for the company and the flexibility to adjust to local requirements. In fact, an EMS potentially provides the basis for accomplishing a great deal more: it can set the foundation for a dynamic interplay between international standards and national rules, including demonstration effects for local companies and assistance in the development of national laws and regulations. The dynamic interplay between international and national standards arises from the fact that firms with operations in several jurisdictions have to meet international standards across jurisdictions as well as national standards within each jurisdiction.

A good example of this process is the IBM ISO 14001 registration. IBM announced that it has received the first edition of a single worldwide ISO 14001 registration that will cover all of the company's global manufacturing and hardware development operations across all of its various business units. The scope of this single registration, awarded by Bureau Veritas Quality International (BVQI), encompasses the company's worldwide EMS at the corporate headquarters level and at 11 of IBM's 28 plant sites involved in the manufacture and development of microelectronics technology, data storage systems, personal systems, servers, and networking hardware. It is IBM's intent that the remaining 17 plant sites will complete the audit and registration process and be brought under the single registration (personal communication from IBM Corporate Environmental Affairs).

Adoption of an EMS that can meet national and international standards inevitably results in adjustments of "international" standards and demonstration effects for national policymakers and local firms. The success of multinational businesses using EMSs to achieve high levels of competitiveness and environmental performance may have a powerful institutional transfer effect for local firms and local environmental regulators. On the other hand, a unified approach across jurisdictions not only enables the company to simplify environmental management across locations but also produces direct benefits in terms of corporate integrity and environmental stewardship.

Finally, EMS standards such as ISO 14001 create a common language (Welford 1996) and a way of thinking about environmental management, which can facilitate communication and partnerships among companies, communities, and governments.

How Does an EMS Relate to Developed Economies and Developing Economies?[2]

Most EMS standards have been created in developed countries for adoption by organizations in developed countries. Examples include the British Standard for Environmental Management Systems (BS 7750), the Canadian Standards Association Standard (CSA 2750), and the European Union Eco-Management and Audit Scheme (EMAS) (Welford 1996). Even ISO 14001, although prepared for global adoption with the participation of developed and developing economies, was largely patterned on existing EMS frameworks in developed countries. To assess how EMSs in general, and ISO 14001 in particular, relate to developing economies, I first examine how these differ from developed economies.

Most developing economies lack effective environmental regulations. Although most of them have already adopted developed country environmental standards and command-and-control means of enforcement, actual enforcement is sporadic because other pressing priorities limit the allocation of financial and human resources to the task or because the authorities do not sufficiently value or account for the existing environmental risks or levels of degradation. In addition, low incomes result in low willingness to pay for environmental improvements and put the environment low on the list of priorities of individuals and governments. At the same time, developing countries tend to be less litigious and less amenable to the command-and-control or "big stick" approach. On the other hand, the use of economic instruments (such as pollution charges and taxes) to internalize environmental costs raises concerns about the loss of competitiveness and the sti-fling of much-needed economic growth to create jobs and alleviate poverty.

In developing countries, firms are much more diverse and heterogeneous in terms of size, ownership, age, and structure than in developed countries. In technical terms, the marginal cost of environmental compliance varies more widely among developing countries than among developed countries. Finally, developing countries experience a much faster process of economic growth, structural change, and turnover of capital and firms, which affects what is desirable and what is achievable (Panayotou and Vincent 1997). Inflexible standards and static sets of laws and regulations soon become obsolete through rapidly changing circumstances.

All these differences between developing and developed countries make EMSs even more relevant because of their flexibility, dynamic nature, use of market incentives, and involvement of shareholders (Ho 1997). Because of the importance of sociocultural factors and the lack of dichotomy between economy and society, a "shareholder" approach to environmental management, accompanied by motivation and inducements, is more likely to suc-ceed than are sanctions, fines, and penalties. Indeed, considerable evidence indicates that informal regulation and community pressure are effective means of improved environmental performance in developing countries that have weak formal regulation, such as in Thailand, India, Indonesia, and China (Huq and Wheeler 1993; Hartman and others 1995; Afsah and others 1996; Pargal and Wheeler 1996; Panayotou and Vincent 1997). Furthermore, the availability of information about environmental performance has been found (by these standards) to further empower shareholders to demand and secure improved environmental performance by firms. Therefore, an EMS that is based on the generation and dissemination of information and the involvement of shareholders has the potential to improve the environment more quickly than an erratically enforced command-and-control, compliance-driven system.

However, if shareholder satisfaction is key to EMSs such as ISO 14001, what does this imply for developing countries whose shareholders include global customers and suppliers (and the global community, in the case of global pollutants such as CO_2 and chlorofluorocarbons [CFCs])? Certainly, the definition of the "shareholder" will be of particular concern to developing countries that shy away from assuming commitments to control CO_2 or feel that their sovereignty is threatened by global community calls for conserving forests and biodiversity as global commons.

At the other end of the spectrum, many developing countries have made environmental improvements at the firm and the national levels but have received little credit for their efforts. Furthermore, developing countries find it difficult to access the green market in developed countries for the green products they produce (for example, "shaded" or "biodiversity-friendly" coffee produced in El Salvador and Costa Rica) because they lack internationally recognized certification. For these developing countries, International Organization for Standardization (ISO) registration would help open access to markets for green products in developed countries.

According to the agreement on Technical Barriers to Trade (TBT) that was in place prior to 1994, rules could be made that differentiate between coffees according to their inherent characteristics (bean type, grind, and so forth). Any such differentiation would be appropriate and would not be a technical barrier to trade. Under the 1994 General Agreement on Tariffs and Trade (GATT) Agreement on Technical Barriers to Trade (WTO 1994), rules could be made about organic coffee, that is, coffee produced by using an environmentally friendly process. These rules would not be presumed to be a technical barrier to trade if the national authorities followed an existing international standard definition for *organic coffee.* Therefore, when ISO adopts the eco-labeling standards, authorities will be able to approve specific ways to label organic coffee that will not be considered technical barriers to trade. With the approval of ISO 14000, authorities could specify that the grower had an EMS based on ISO 14000. Such rules on eco-labels and EMSs related to coffee would be consistent with the new requirements of the TBT agreement (Krut and Gleckman 1998).

How Do EMSs Transcend Nation-States?

In a world where even parts of simple products (such as textiles and garments) are produced in different jurisdictions and insurance companies spread risk across nations, the implementation of EMSs will, by necessity, transcend nation-states. However, rather than being a constraint, it is instead an opportunity to leapfrog state bureaucracy by using free enterprise devices (such as supply chains, insurance pools, and global funds) to improve environmental performance through the adoption of an EMS driven by shareholder satisfaction. After a firm adopts an EMS such as ISO 14001, it may choose to require all its suppliers, investors, lenders, and insurers to conform to its environmental policy goals, although few EMSs actually have achieved this level of integration thus far. In this way, implementation of a management system based on ISO 14001 may become a prerequisite for doing business: customers may demand that suppliers meet specific environmental standards and have ISO certification.

With customers, suppliers, investors, and insurers spanning the globe, adoption by a substantial number of firms of EMSs based on the ISO standard would create a worldwide network of environmental performance commitments that transcend nation-states and flow in the same channels and directions of daily business transactions. The dynamics and multiplier effects of this interface are difficult to fathom, because the customer of one company is the supplier of another. At the same time, bankers—who play a key role in moving money across borders—are sensitive to aggregate environmental liabilities of firms to which they lend, and they influence the environmental behavior of firms through credit rating and the cost of capital. For example, Chilgener, a Chilean firm, lost 5% of its market value in April 1992 after it released a cloud of toxic air pollution over Santiago, Chile. Its value rebounded in September 1992 after it announced an investment of $115 million to control air pollution (Dasgupta and others 1997). Dasgupta, Laplante, and Mamingi (1997) found that companies with improved corporate environmental practices were able to increase their shareholder value by up to 5% and to attract more investment because of a perceived reduction in risk after the adoption of EMSs. A perceived reduction in environmental risk also may result in lower insurance premiums. Because insurers pool risks across companies and across borders, the adoption of an EMS such as ISO 14001 by a firm in one country affects other firms across national borders via the common insurance pool.

Which Attributes of an EMS Deliver Greater Global Value?

The single, most important attribute of EMSs (in terms of global value generation) is flexibility. Compliance costs differ widely within and among developing and developed countries. The flexibility of an EMS allows each company to adopt different goals and priorities and to pursue them at a different pace, generating substantial cost savings in terms of compliance cost, helping to preserve competitiveness, and ensuring that some environmental improvement occurs (and will continue to occur) in developed and developing economies.

A related feature is that to the extent that an EMS (in combination with other measures) helps preempt the use of the environment as a nontariff trade barrier, the incremental gains from trade—which otherwise would not have materialized without the EMS adoption—can be considered part of the global value generated by EMSs.

Standardization of the process of setting goals and priorities, assigning responsibilities, and measuring and reporting results is another feature of EMSs that contributes greatly to their global value, because they standardize the process without artificially imposing uniform performance standards. This feature actually acts as a break in the race for the bottom and, in fact, reverses this process from backsliding to ratcheting up and moving forward. Standardization, such as might occur in an EMS based on ISO 14001, also saves considerable resources for firms that operate across many jurisdictions. An equally important feature of an EMS, in terms of global value delivery, is the requirement that claims of improved environmental performance (progress toward the goals and targets) be regularly audited and externally verified. This feature ensures consistency among enterprises as well as credibility of the process. The voluntary nature of EMSs is both a weakness and a strength. It is a weakness because the worst polluters may choose not to participate; it is a strength because it relies on motivation rather than coercion and allows each firm to participate at its own pace.

Which Attributes of an EMS Have Less Standing with Different Stakeholders?

The voluntary nature of an EMS is its single most undesirable attribute among environmentalists, who decry the lack of assurance that ISO 14001 would improve environmental performance. The fact that an EMS is a process, not a performance standard, further compounds the uncertainty of environmental outcomes. Even if all countries and firms are ISO 14001–certified, there is no guarantee that the environmental performance of firms would rise. A related concern is that ISO 14001 has missed opportunities to set global environmental standards for sustainable industrial development. According to Krut and Gleckman (1998), "there is a strong possibility that ISO 14001 cannot achieve sustainable development or even environmental performance improvement—for a certifying firm, for its suppliers, for regulators ... for the intergovernmental community seeking to address global environmental impact of industrial facilities in their neighborhoods...."

For these reasons, Krut and Gleckman (1998) propose an ISO-plus approach, where the three "plus" elements include key aspects of sustainable industrial development: transparency, democracy, accountability, and partnership; application of international environmental standards; and best available practices.

Whereas it is difficult to argue against more transparency and accountability, the demand for common and binding international environmental standards and enforcement of "best practices," regardless of local conditions, is misguided and could eliminate most of the beneficial features of an EMS (flexibility, gradual progress toward the goals, and voluntary commitments). Another set of criticisms involves the exclusiveness of the ISO decision-making process and the fact that standards officials from developing countries came into the ISO 14001 process very late in the game (after the summer of 1995), when ISO 14001 was approved (Table 5-2). There has been a sense that formal procedures "kept developing countries, [nongovernmental organizations], and others out of the ISO drafting and negotiation process" (Krut and Gleckman 1998).

A third criticism concerns the relationship between uptake and implementation of management systems based on ISO 14001 and WTO. Under

Table 5-2. Developing Countries' Participation in the Formulation of ISO 14001

Region	Number of countries
Latin America and the Caribbean	14
North Africa	5
Other Africa	7
West and central Asia	7
Eastern and southern Asia	15
Developing Europe	1
Total	49

Source: Krut and Gleckman 1998.

GATT rules, ISO and other private-sector organizations have the authority to set international standards that have legal standing under the 1994 GATT Agreement on Technical Barriers to Trade. As such, ISO standards benefit from GATT sanctions: governments that set national standards can ignore ISO standards only at their own peril, because standards that differ from ISO standards may be challenged by foreign governments as technical barriers to trade. Thus, some would assert that ISO standards may act as obstacles to developing innovative approaches to environmental management. Again, this criticism is somewhat misplaced and exaggerated because ISO 14001 establishes a framework for a management system, rather than specifying particular performance standards.

A related concern is that ISO 14001 certification by a firm is a ticket to market access, but the costs of achieving certification are much higher for firms in developing economies than in developed economies because developing country governments and firms lack experience with corporate environmental management. Leapfrogging the command-and-control phase of regulation to go directly to ISO 14001 may be risky for environment and public policy (Krut and Gleckman 1998). The concern about lack of experience and limited capacity for corporate environmental management is understandable, but it can be addressed through special arrangements and capacity-building assistance. This concern in no way justifies the time and resources wasted on inflexible command-and-control regulations, which developing countries can ill afford before they can benefit from the flexibility and dynamic incentives of EMSs (IISD 1996).

The final criticism involves the formal structure of most EMS standards, including ISO 14001. The concern is that an EMS framework may be too bureaucratic: it requires collecting and storing large amounts of information; keeping detailed records; conducting surveys, reviews, and audits; and following several formal procedures that may be as intrusive as government interventions and may detract as much from the firm's main mission.

Critics claim that regulatory relief is bought at a high bureaucratic price. Although there is an element of truth to this criticism, it is more of an issue in small and medium-sized enterprises (SMEs), which place the environment low on their list of priorities and lack the scale and financial resources to have a full-blown, externally certified EMS. Worldwide, SMEs account for about 70% of national product in developing countries, making this concern a major one that must be addressed through an abridged or simplified yet effective EMS for SMEs.

What Attributes of an EMS Merit Development and Replication?

In 1995, Global Environmental Management Systems conducted a survey of top executives from 11 U.S. corporations (including IBM, Procter and Gamble, 3M, Motorola, and General Electric) to identify the strengths and weaknesses of implementing an EMS. The respondents also were asked to recommend what could be done to increase the acceptance of ISO 14001 EMSs by industry.

The results of the survey indicated that the major strengths of EMSs included the facilitation of trade as a result of the prevention of nontariff barriers (31%), the enhancement of quality management through harmonization and benchmarking of improvements (25%), and the reduction of liability and lower insurance premiums (14%). The major weaknesses were the costs of implementing an EMS (28%), the lack of specific environmental performance criteria or standards (22%), and the lack of public awareness necessary to encourage EMS acceptance. Respondents suggested that the most effective ways to improve the acceptance of EMSs based on ISO 14001 would be to provide stronger regulatory incentives (58%), strengthen its trade benefits (18%), and increase public awareness (12%).

Indeed, two areas that warrant further development are strengthening the regulation incentives for adoption of an EMS and creating enhanced measures of performance. Toward this end, the Environmental Leadership Program (ELP) of the U.S. Environmental Protections Agency (EPA), which aims to encourage voluntary compliance and build working relationships with shareholders through the adoption of an EMS such as ISO 14001, has added compliance assurance, employee involvement requirements, and community outreach elements as well as environmental performance measures and incentives for self-policing. Wisconsin's Department of Natural Resources, on the other hand, has a created an alternative regulatory track that incorporates a three-tiered model of fines for criminal and civil penalties, in which the lowest fines are imposed on companies that are ISO 14001–registered. Similarly, Pennsylvania has indicated that fines, penalties, and surveillance schedules may be mitigated if an organization uses an EMS.

In Australia, companies that violate pollution laws may be exonerated if they can demonstrate due diligence, that is, that they took reasonable steps to avoid the offense. Having an ISO 14001 EMS registration or some other effective EMS in place may go a long way toward proving due diligence.

Another incentive is a green procurement policy, whereby the government gives preference to ISO 14001–registered suppliers. Similarly, countries may require or favor companies with ISO 14001 registration in their foreign investment policies.

What Research Questions Will Help Deploy an EMS on a Global Scale?

ISO 14001, the international EMS standard, is still far from being globally adopted. Developed countries lead the adoption; by January 1999, Japan led them with 1,397 industries certified (Table 5-3). Developing countries are far behind (Table 5-2). A major obstacle to the deployment of EMSs on a global scale is the lack of documented evidence of their environmental performance and their effects on competitiveness. To obtain such evidence, it is necessary to compare the environmental and financial performance of firms with and without (not before and after) EMSs. This task involves constructing a business-as-usual scenario (baseline) and evaluating the EMS's environmental and financial outcomes against such a scenario. Alternatively, one may attempt a regression analysis of environmental outcomes for firms against a set of determinants that include the presence or absence of an EMS with different attributes to evaluate the effects of individual attributes. To carry out such an analysis, data are needed on the environmental and financial performance of a large sample of firms and their determinants (characteristics or descriptions of the industry, the firms, the technology, the regulatory framework, the shareholders, and the EMS). Designing better EMSs must be part of such an exercise. Therefore, data also must be collected on the design criteria used to create the architecture of various EMSs and their attributes, such as degree of flexibility, regulatory incentives, public environmental awareness, community outreach, and compliance assurance.

Furthermore, the effectiveness of EMS programs based on ISO 14001 in developing countries with uneven enforcement and a strong focus on economic performance must be tested. In such a case, the role of the local community and other civil society stakeholders may be key in bringing about environmental improvement through an EMS. Cross-country regressions of developed and developing countries with and without an EMS would help assess the role played by conditions in developing countries, such as rapid growth and structural change. The effect of an EMS on competitiveness and

Table 5-3. ISO 14001 Certification in Japan by Industry Sector

Sector	Certification (%)
Electric machinery	46.4
General machinery	10.2
Chemical technology	8.3
Transport equipment	6.0
Precision instruments	5.5
Services	2.9
General construction	2.4
Iron and steel	1.9
Cement products	1.7
Rubber products	1.5
Oil products	1.3
Other	11.9

Source: Japanese Standards Association 1999.

environmental improvement needs to be assessed if an EMS is to have any chance of being adopted by developing countries.

In terms of EMS applications to improve global environmental performance, prospects may exist with regard to the participation of developing countries in global efforts to control greenhouse gases through the clean development mechanism. For example, adopting an EMS may help establish consistent and comparable carbon emissions baselines across countries.

However, before EMSs will have a fair chance of adoption on a global scale, research on the following issues is needed.

■ How do EMSs affect the environmental and financial performance of firms compared with a business-as-usual scenario?

■ How does the performance of EMSs compare in integrated (or industrial) and emerging economies? Can we identify the role played by the level of economic integration and development in the effectiveness of EMSs in creating global value?

■ What is the optimal mix of standardization and flexibility that would maximize global value of EMSs? Is an imperfect ISO 14001 that is common to all countries, industries, or firms superior to more fine-tuned EMSs that are tailored to individual countries, industries, or firms?

■ What design criteria should be incorporated into the architecture of various EMSs with the desired attributes of flexibility, standardization, and compliance assessment?

■ What is the potential applicability and value of EMSs in the implementation of global environmental conventions, including the Kyoto Protocol and the Convention on Biological Diversity?

Conclusion

In a world of growing environmental awareness and an increasingly global economy with free flow of goods and capital, common environmental standards are inevitable. Yet with such a diversity of conditions, costs, technology, and preferences across firms, industries, and countries, no single set of environmental standards can be appropriate if the aim is to achieve improved environmental performance without unduly limiting economic competitiveness and growth. EMSs, when appropriately designed and implemented, promise to advance economic and environmental performance by taking due account of individual firm, industry, and national conditions and optimizing between standardization and flexibility.

Acknowledgements

Partial support from the AVINA Group grant to the Center for International Development is gratefully acknowledged.

The author is indebted to Suzanne Dickerson of BMW Group, Darlene Pearson of the North American Commission for Environmental Cooperation, and Edward Quevedo of Pillsbury Madison & Sutro LLP for very helpful comments and suggestions on an earlier draft.

Notes

[1]"Before the creation of ISO 14000 as an international standard in 1996, countries such as the U.S., France, Ireland, the Netherlands, and Spain had developed their own EMS standards. The possibility that these diverse EMS frameworks could result in barriers to international trade led to heightened interest in formulating an international voluntary standard for EMSs" (U.S. EPA 1998).

[2]Although I follow the convention of distinguishing between "developed" and "developing" economies to aid communication, a more appropriate distinction for our purposes would be "integrated" and "emerging" economies, with the latter including the transitional economies of Eastern Europe and the former Soviet Union.

References

Afsah, S., B. Laplante, and D. Wheeler. 1996. Controlling Industrial Pollution: A New Paradigm. Policy Research working paper 1672. Washington, DC: World Bank.

Alexander Grant & Associates. Various years. Surveys of industry associations. London, U.K., and Nairobi, Kenya: Alexander Grant and Associates, Certified Public Accountants and Management Consultants.

Dasgupta, S., B. Laplante, and N. Mamingi. 1997. *Pollution and Capital Markets in Developing Countries.* Development Research Group. Washington, DC: World Bank.

Epping, M. 1986. Tradition in Transition: The Emergence of New Categories in Plant Location. *Arkansas Business and Economic Review* 19(3): 16–25.

Fortune Magazine. 1977. Various issues.

Hartman, R., M. Huq, and D. Wheeler. 1995. Why Paper Mills Clean Up: Determinants of Pollution Abatement in Four Asian Countries. Working paper. Washington, DC: World Bank.

Ho, F. 1997. Catching Up and Converse. In *Global Competitiveness Report 1997.* Geneva, Switzerland: World Economic Forum.

Huq, M., and D. Wheeler. 1993. Pollution Reduction without Formal Regulation: Evidence from Bangladesh. Working paper 1993-39. Washington, DC: World Bank.

IISD (International Institute for Sustainable Development). 1996. *Global Green Standards: ISO 14000 and Sustainable Development.* Winnipeg, Manitoba, Canada: IISD.

Japanese Standards Association. 1999. *International Environmental Systems Update.* Akasaka, Japan: Japanese Standards Association. http://www.jsa.or.jp (accessed September 1999).

Krut, R., and H. Gleckman. 1998. *ISO 14001: A Missed Opportunity for Sustainable Global Industrial Development.* London, U.K.: Earthscan.

Lyne, J. 1990. Service Taxes, International Site Selection, and the "Green" Movement Dominate Executives' Political Focus. *Site Selection* October.

Panayotou, T., and J.R. Vincent. 1997. Environmental Regulation and Competitiveness. In *Global Competitiveness Report 1997.* Geneva, Switzerland: World Economic Forum.

Pargal, S., and D. Wheeler. 1996. Informal Regulation of Industrial Pollution in Developing Countries: Evidence from Indonesia. *Journal of Political Economy* 104(6): 1314–1327.

Schmenner, R. 1982. *Making Business Location Decisions.* Englewood Cliffs, NJ: Prentice-Hall.

Stafford, H.A. 1985. Environmental Protection and Industrial Location. *Annals of the Association of American Geographers* 75(2): 227–240.

U.S. EPA (Environmental Protection Agency). 1998. EPA Position Statement on Environmental Management Systems and ISO 14001 and a Request for Comments on the Nature of the Data to Be Collected from Environmental Management System/ISO 14001. *Federal Register* 63(48): 12094–12097.

Welford, R. (ed.). 1996. *Corporate Environmental Management.* London, U.K.: Earthscan.

Welford, R., and A. Gouldson. 1993. *Environmental Management and Business Strategy.* London, U.K.: Pitman Publishing.

Wilson, K. 1996. Benefit of Environmental Management and ISO 14000. *Transformation Strategies.* http://www.clickit.com/newsweb/nwsrch.htm (accessed September 1999).

Wintner, L. 1982. *Urban Plant Siting.* New York: Conference Board.

WTO (World Trade Organization). 1994. Global Agreement of Tariffs and Trade (GATT), Agreement on Technical Barriers to Trade. In *Final Act of the Uruguay Round* (Annex 1A: Multilateral Agreements on Trade in Goods). Geneva, Switzerland: World Trade Organization.

Part 2

Implications for Public Policy

Environmental regulatory systems typically emphasize compliance with a set of rules designed to maintain clean air, water, and land. Yet many facilities are able to do a better job managing their environmental impacts than is required as a matter of law. Already, many organizations are inventing new ways to take responsibility for the unintended side effects of their products and processes. For some, the adoption of an environmental management system (EMS) is an important component of these efforts. How can regulation build on this momentum? What role, if any, should EMSs play in a new environmental policy agenda?

In Part 2, the authors present diverse answers to this question. In Chapter 6, William R. Moomaw argues that there are no "environmental" problems per se. The problems of air and water pollution, hazardous waste sites, and climate change are symptoms of a mismatch between how human beings meet their needs and the physical and biological workings of the natural world. Most EMSs are adopted to address environmental problems, not the underlying social problems that cause this mismatch. Moomaw argues that EMSs should contribute social value that encompasses equity, justice, democratic participation, security, education, and trust. He urges policymakers to broaden their conceptions about the goals of EMSs and begin to consider what he calls "sustainability management systems."

Moomaw presents three case studies of industries and firms that have attempted to shift from a compliance stance to business principles based on sustainability. The first case focuses on firms in the chemical industry that have used a structured EMS approach. Although these firms succeeded in improving environmental quality, Moomaw argues that they failed to enhance other social values. In the second and third cases, firms did not implement formal EMSs but instead implemented practices on an ad hoc basis in response to particular needs. The specific practices they selected— such as building a school for workers' children—created social value for and mutual trust between the firm and the surrounding community.

In Chapter 7, Shelley Metzenbaum raises questions about the role of EMSs in environmental regulation. Many people from business, government, and universities now argue for a move toward a more performance-based and information-driven regulatory system. Under the most optimistic scenario, broad use of EMSs could lead to a radical redefinition of the role of agencies. Instead of relying on technology-based standards, agencies could rely on a facility's EMS to generate performance-based environmental goals. Instead of monitoring the facility to verify that managers were in compliance with government standards, regulators could rely on reports prepared by third-party auditors certifying that firms had established and met their own performance targets. Current EMS standards, such as ISO 14001 and Responsible Care, do not encourage public disclosure of environmental performance information or establish a system of environmental performance metrics to use to compare different firms. Metzenbaum concludes that these EMS standards therefore may contribute little to the performance-focused system she envisions.

In Chapter 8, Cary Coglianese brings up a question central to the themes discussed throughout this book: do EMSs yield improved results? He argues that in principle, an EMS may "draw in" employees, signaling that they should make environmental performance a priority. Once established in an organization, an EMS may "lock" managers into progressive improvements. However, he also suggests that factors independent of the management system (such as managers' overall commitment) may better explain a facility's environmental performance. Agencies should be aware that policies that aggressively promote EMSs could fail to support or even stand in the way of earnest efforts on the part of managers to address environmental problems.

If we assume that EMSs do yield better results, what steps could agencies take to encourage their use? Coglianese answers this question by reviewing policy options that could lower the costs of EMSs, increase their benefits, or require their use. He concludes that technical assistance, audit protection, enforcement forbearance, and private mandates appear the most promising—although still limited—options to pursue. Each of these options offers at least a moderate incentive to firms to improve their environmental management and are legally and politically acceptable. In contrast, the promise of regulatory flexibility, as offered in some of the performance-based programs being developed at the state and federal levels, will drive up the costs of participation for facilities managers because agencies probably will require extensive documentation and negotiation to determine who deserves this form of special treatment.

In Chapter 9, Jerry Speir reviews the experience of 11 states that have developed tiered regulatory systems to provide benefits to firms that meet specified performance criteria. He offers numerous examples of instances in which EMS implementation has led to measurable reductions in the environmental impacts of the facilities. However, determining the improvements attributable to EMSs, he argues, is only half of the public policy equation. The other half requires policymakers to ask what kind of benefits should be offered to firms that implement EMSs. Speir concludes that such determinations are political questions that need to be answered, in the final analysis, through the democratic process.

The common theme in Part 2 is that policymakers should be cautious about creating a system that rewards firms solely for EMS adoption. Managers can contribute social value and even steer a course toward sustainability without implementing a formal EMS. At this point, managers appear not to be using EMSs to generate and disclose information about their performance to the public. As a result, EMSs appear to contribute little to building public trust and are not yet yielding information that could be used in a performance-focused, information-driven regulatory system. Furthermore, implementation of a government system to recognize and reward firms with certain kinds of EMSs presents difficult questions of how much "extra" a firm needs to provide to gain entry into the privileged tier, how much agencies should offer in return, and how any such deals should be monitored.

6

Expanding the Concept of Environmental Management Systems to Meet Multiple Social Goals

William R. Moomaw

Imagine a world in which every company, utility, municipal government, household, and even university had an environmental management system (EMS) in place. Every action taken by the firm was subject to a financial and an environmental assessment before any action was taken. It sounds like a plan to create massive unemployment for thousands of state and federal environmental enforcement officials. After all, what would they do if such an EMS world existed?

For better or for worse, environmental officials need not worry. Aside from the unlikely prospect of such a world, there is obviously still a need to match management practices to implementation to verify that environmental quality requirements are met. Three major constituencies are likely to oppose such a shift: public interest and environmental groups, who are distrustful; regulators, who prefer control; and companies themselves, which prefer certainty. Yet, profound evidence suggests that the current environmental regulatory system, having brought us a substantial distance from the polluted world of the past, is in need of a major overhaul.

George Meyer, former secretary of the Wisconsin Department of Natural Resources, makes a compelling case for a new approach to environmental protection. He argues that the current control tier of the regulatory process needs to be supplemented by a "green tier" to address the estimated 80% of environmental needs that currently are not being addressed. His vision would replace reliance on strict command-and-control regulations with a more flexible, EMS-based approach that involves incentive-based contracts between the regulator and the regulated. Such a system has been implemented in the Netherlands with some notable successes (Meyer 1999).

The EMS approach, which is a major step forward, nevertheless is still premised on the notion that we are dealing with pollution problems. This perception may be getting in the way of an effective means for action. I suggest that there are *no environmental problems.* Air and water pollution; solid and hazardous waste disposal; and even biodiversity loss, acid rain, and climate change are not problems but rather symptoms of the mismatch between how individuals and society meet our needs and wants on one hand and the physical and biological processes of the planet on the other. Unless this fundamental shift in mind-set is made, we will be limited in what can be accomplished, and our actions will be equivalent to using aspirin to relieve headache pain from a brain tumor.

The social invention that has been created by human society to provide and allocate needs and wants is the economic system, which in turn is supported by the legal and political structure of the government that defines it. If a better system than traditional environmental regulation is to be found, it must reach the root causes and not merely the environmental symptoms. Aligning economic and environmental incentives lies at the heart of any alternative regulatory structure. Turning around the legal structure to make such a change possible will be a major challenge for the political system.

Secondly, for a management system to function, it must adequately address the social concerns of society. What is missing from most EMSs to date is a component or metric of social value. For members of society, the heart of the matter is social value in the guise of environmental performance. However, the public is searching for additional social values that usually are not stated explicitly. In addition to environmental quality, these social values include safety and health; equity, fairness, and justice; community values and participation; security; education; and most important, trust.

In this paper, I describe three case studies of companies that have initiated imaginative EMSs. I interviewed top officials of all three companies and visited one of the three corporate sites. I also try to identify the social dimensions that must be a part of future corporate practices if EMSs are to become an accepted part of the political and economic system of production.

Three Case Studies

The case studies represent three very different kinds of companies and EMSs. The first examines the approach of large multinational chemical companies that have been moving beyond traditional environmental compliance measures in their efforts. The second case is a U.S. start-up mining operation in Montana. The third example is a small Dutch-owned company that produces cut flowers in Kenya for export to European and Japanese markets.

Case 1: Multinational Chemical Companies

The chemical industry provides some of the most dramatic examples of a change in environmental management practices. Long noted as a principal polluter of air and water at and beyond the plant boundary, the industry was transformed by the environmental regulatory revolution and the oil price shocks of the 1970s. Several acute crises forced the industry to come to terms with its own operation and with the growing dissatisfaction of society at large with its performance.

Rachel Carson's (1962) book *Silent Spring* defined the terms of the debate over chemicals and their release during manufacture and use. Her account of how DDT, a pesticide then widely used in food production, concentrated in mother's milk and also led to the near extinction of bald eagles and other birds created a public outcry over chemical products in the environment. Allied Chemical Corporation's mismanagement and attempted cover-up of the extensive contamination of the James River in Virginia by Kepone, another pesticide, and the tragic mercury poisoning of villagers in Minamata, Japan, from effluents of Chiso Chemical Company supported the view that the chemical industry was irresponsible. A drumbeat of reports of illegal chemical dumping in New Jersey, the contamination of homes and a school at Love Canal in New York, and the contamination of drinking supply wells in Woburn, Massachusetts, produced more than "a civil action." By the mid-1980s, the industry's public credibility had hit rock bottom. The catastrophic release of methylisocyanate (MIC) from a Union Carbide chemical plant in Bhopal, India, was the shock that brought real change to the industry as a whole. The accident killed more than 2,000 people and seriously injured tens of thousands more.

I was chair of the American Chemical Society's Task Force on the Toxic Substances Control Act during the 1980s. The group consisted of half a dozen chemical company scientist–administrators, plus a couple of university and government representatives. Before Bhopal, the discussion revolved around how unreasonable it was to regulate the chemical industry and how the process might be made less onerous; the public and the regulators simply needed to be educated about the general safety of chemicals, responsible management by the companies, and the benefits that chemicals provided the public. As one advertising slogan put it, "Without chemicals, life itself would be impossible!"

The meeting we held one month after the Bhopal tragedy was entirely different. All of the industry representatives were clearly shaken. Yes, the unthinkable could happen, and a smaller release of MIC occurred soon thereafter—from an American facility. The tone of the discussion changed abruptly. The Chemical Manufacturers Association (CMA), now the American Chemistry Council (ACC), accelerated its Responsible Care program to raise the environmental and safety standards of an entire industry. Many features of today's EMSs were incorporated into this early effort.

Another early example of a practical EMS was the establishment of the Pollution Prevention Pays (PPP) program at 3M Corporation in the 1970s by James Ling. Driven by a combination of regulatory and cost concerns, the program wisely succeeded in aligning environmental and economic interests. Other chemical firms made similar shifts during the 1980s and '90s. Dow Chemical, which suffered under the cloud of having manufactured napalm and Agent Orange during the Vietnam War, once challenged the U.S. Environmental Protection Agency (EPA) in court to exempt monochloro-biphenyl from a ban of polychlorinated biphenyls (PCBs) on the technical ground that "poly" (many) does not include "mono" (one). Like many companies at the time, Dow adopted an aggressive and defiant position toward the public and government. However, under new management, it later instituted its own highly successful Waste Reduction Always Pays (WRAP) program. Dow not only dramatically reduced its chemical releases into the environment through its innovative pollution prevention programs but also pioneered chemical waste management systems. The company also became a partner in the world's largest industrial combined heat and power project with its local utility and developed innovative programs in industrial energy efficiency. The company gained such respect for its environmental initiatives that David Buzzelli, a senior vice president, was selected to be co-chair of the President's Council on Sustainable Development.

At the time that stratospheric ozone depletion was first identified as a potential environmental problem in 1974, DuPont was responsible for one-quarter of the world market of the chlorofluorocarbon (CFC) chemicals that were implicated as the prime agent of ozone destruction. In 1989, Edgar Woolard Jr., the new CEO, created a major stir by declaring that DuPont would use a common environmental standard for all of its facilities globally. He also announced that instead of battling the regulators and the public on technical grounds, the company would from then on consider the desires of society with respect to safety and the environment in its corporate decision-making. A few months later, DuPont dramatically announced that it would halt CFC manufacture worldwide by 2000. It later moved the phase-out date to 1995 and had to be persuaded by Vice President Al Gore to continue manufacturing for one additional year to meet the needs of user firms. DuPont clearly was ahead of not only its competitors but also government regulators and diplomats, who at the time still assumed that CFC production would only be halved for industrial countries by 2000.

DuPont has continued to take dramatic public action. In September 1999, Dennis H. Reilley, executive vice president and chief operating officer of DuPont, announced an ambitious three-point goal to address climate change by 2010 (Reilley 1999):

- reduce global carbon-equivalent greenhouse gas emissions by 65%, using 1990 as the base year;
- hold total energy use constant, using 1990 as a base year; and
- use renewable resources for 10% of global energy use.

In making its announcement, DuPont joined the ranks of a handful of major multinational corporations taking action to address climate change. These companies include BP-Amoco, which has committed to a 10% reduction in carbon dioxide emissions worldwide by 2010, and Shell, which has committed to slightly lower reductions of greenhouse gases within a shorter time frame.

Behind the scenes, DuPont was shifting the basis of its operating principles from environmental management to sustainable development. Doing more with less and establishing congruent reward systems for managers to reinforce sustainability initiatives became organizing principles in several DuPont divisions. Shifting the focus of polyethylene production from emphasizing tons of product sold to providing thinner plastic sheets substantially reduced raw material use, associated pollutant emissions, and energy consumption and raised profits. Instead of trying to sell more auto paint to car manufacturers, DuPont created a new technology for vehicle painting and a business that provided painting as a service, using far less paint than with old technology while creating less pollution and increasing profits (personal communication from Dawn Rittenhouse at The Industrial Ecology Gordon Conference, Colby-Sawyer College, New London, NH, June, 10, 1998). Furthermore, the company has set the goal of reducing its environmental footprint by obtaining one-quarter of its revenues from renewable resources by 2010 (Reisch 2000).

DuPont's goal is to have environmental and sustainability concerns permeate the entire organization, just as safety has been a hallmark for a company originally founded as a munitions and explosives manufacturer. In fact, DuPont sees sustainable innovations as the major opportunity for expanding market share and ensuring growth in excess of the gross domestic product (personal communication from Mack MacFarland, September 12, 1999, Lillihammer, Norway).

The Responsible Care program of the ACC is a formal EMS that requires members to commit to specific practices and procedures. Unlike ISO 14000, Responsible Care commits its members to compliance with environmental laws and regulations, requires a degree of transparency in dealing with regulators and the public, and imposes sanctions that include expulsion from the program for falsifying reports and other transgressions. Responsible Care called on companies to set performance goals only recently, but these goals are still voluntary, and no standard set of common goals exists for the 190 member companies and 60 affiliates. Similarly, third-party verification of management systems, introduced in 1996, remains voluntary, and less than one-half of the member companies had implemented it by mid-2000 (Reisch 2000). The status of EMSs within chemical firms is less uniform. Those that adhere to Responsible Care or ISO 14000 guidelines have formal documents, and companies such as 3M and Dow long have had specific pollution prevention guidelines. The DuPont system appears to be a collection of general directives, departmental and divisional procedures, and specific action plans on high-profile issues such as ozone depletion and climate change. A single, comprehensive, corporate-wide EMS may be more the exception than the rule.

Case 2: Stillwater Mining

Stillwater Mining was formed in the early 1990s as a spin-off from a large energy company. The mine is located north of the Wyoming border on federal land in Montana. The principal output of the mine is rare metals such as platinum and palladium, which are used as catalysts in industrial processes, in catalytic converters in vehicles, and potentially in fuel cells as a next-generation source of clean electricity and automobile energy. The only other significant sources of these important metals are in South Africa and Siberia.

Developing a new mine near Yellowstone Park at the same time that a highly controversial Canadian-backed gold mine was being proposed proved a challenge. The founding CEO of Stillwater, Charles Engles, is environmentally aware and technically trained. He also had a keen desire to avoid the long contentious process that was hampering other projects in the region. In addition to wanting to develop the mine on an environmentally sound basis, it is clear that he had an excellent understanding of several important principles of negotiation.

His first goal was to avoid the early development of hard positions by any of the parties, knowing that such positions become increasingly difficult to change as negotiations proceed. Second, he insisted on meeting with and involving all stakeholders from the beginning of the process. Third, he instructed his management team to take all of the parties' concerns seriously and to try to anticipate and respond to their concerns with actions by the company. Finally, even after gaining the upper hand, he engaged all parties throughout the process.

The stakeholders included officials of the U.S. Forest Service and Bureau of Land Management, EPA, state environmental and economic development officials in Montana, local ranchers and citizens, and grassroots environmental activists. Despite their active opposition to the gold mine and the potential environmental benefits of a new, inexpensive source of platinum and palladium for auto catalytic converters and fuel cells, national environmental organizations played almost no role in the process. Attempts to interact with Congress over pending legislation to change American mining laws were unsuccessful.

Realizing that it would be necessary to address the concerns of federal and state officials as well as local environmental activists, the company immediately established an environmental management process. The CEO made it clear to all of his staff from the beginning that development of the mine operation would conform to all environmental laws and regulations and that shortcuts would not be tolerated. Furthermore, he insisted that the company be as transparent as possible in its dealings with both officials and the public.

According to Engles, it was relatively easy for the mining operation to satisfy state and federal regulatory requirements, and the company had little difficulty in doing so. The project had two inherent advantages. First, it is an underground mine with a small surface footprint. Mining debris is highly inert, and most of it is readily disposed of in the mined-out portions of the underground excavation. Second, although close to Yellowstone Park, the mine lies outside the drainage basin of the park. Nevertheless, local grassroots environmental groups opposed the project at hearings and in court. Many felt that they had a mission to keep the mine from being developed.

The management strategy of the company was to take each objection and devise a response to it. Erosion from roads was dealt with by going well beyond U.S. Forest Service requirements. One local group argued that nitrogen-containing explosives being used might lead to nitrogen-based eutrophication of local streams. Even though measurements and analysis demonstrated that any increase in surface water nitrate was too small to detect and no government regulators required the company to address the problem, Stillwater agreed to have a contingency system in place should it become necessary. These and other actions were taken even after the company had won decisions in court. This patient approach and the fact that the company had anticipated some environmental concerns won Stillwater Mining a degree of cooperation from some if not all of its local adversaries. The project also won a regional environmental award for its efforts, and the mine is now in production.

In an interview, Engles stated that at no time was a formal EMS in place for the company. Instead, a set of environmental and procedural principles and specific environmental management practices were introduced in response to issues as they arose. He and the appropriate managers and decisionmakers met and decided what should be done in response to a specific potential problem identified from within the firm or when a challenge came from a local environmental group or an official. To date, the company has no written EMS, but according to Engles, the company remains committed to avoiding environmental damage and contamination of the site, and environmental considerations are integral to all management and technical decisionmaking (personal communication from Charles Engles, September 19, 1999, Lewa, Kenya).

Case 3: Oserian Development Company, Ltd.

Located near the shores of Lake Naivasha in Kenya, Oserian Development Company, Ltd., was started in 1992 by two Dutch brothers who still own the company. By 1999, the firm was producing and exporting 250 million cut flower stems per year to supermarkets and Tele-Flora in the Netherlands and Britain and to a growing group of customers in Japan.

Lake Naivasha is the second largest freshwater lake in Kenya. Its importance for water bird migration and habitat is recognized by having been designated a site of special significance under the international Ramsar Convention by the Kenyan government (discussed later). Because of the availability of free lake water for irrigation, the region has become the site of a large agricultural industry of flower and vegetable growing for export during the past decade. Ten companies now employ more than 30,000 Kenyans in the region, and the industry is a rapidly growing earner of foreign currency.

Oserian, which is the Masai word for "peaceful place," is situated between Hell's Gate National Park and Lake Naivasha ("turbulent waters"). The company owns 5,000 acres, including more than four miles of lakeshore. Its fields, greenhouses, and reservoirs cover more than one-half of the site, and new fields are being developed. Facilities include a major processing center and a warehouse system, where the flowers are sorted, graded, and packaged as bouquets and bunches ready for retail markets. A fleet of company-owned refrigerated trucks transports enough flowers each week to fill 15 jumbo jets that fly from Nairobi (100 miles from Lake Naivasha by a poor-quality highway) to Amsterdam, the Netherlands (a nine-hour flight). After passing through the international flower market at Aalsmeer, just south of the Amsterdam airport, the flowers are in supermarkets and delivered to individual customers within 24 hours of the time that they were picked. A small but growing number of orders are transshipped from Amsterdam to Tokyo, Japan, where they are in the market within two days of being cut and sorted in Kenya. Kenya is now the second largest exporter of flowers to the Netherlands behind Israel.

In September 1999, I was able to visit the Kenyan site and tour it with Tom Fraser, the Oserian production director, and several of his field and greenhouse managers. An official from the growers' association accompanied us as part of his inspection effort to ensure compliance with agreed-upon environmental and labor practices.

The environmental and social issues facing a company like Oserian are formidable. Kenya not only is a poor country but also has been plagued by wide-scale corruption and mismanagement. Until the 1990s, the population was estimated to be growing at an annual rate of 4% (doubling every 17.5 years). This rate has now declined to 2.9% (doubling every 24 years). The fertility rate has dropped in half, from six children per woman to three. Major problems include AIDS (acquired immune deficiency syndrome), malaria, and multiple tropical diseases as well as an exceptionally high rate of automobile accidents and urban crime. Although all Kenyans are entitled to healthcare, available funds are only $0.07 per person per year. Intertribal rivalry has prevented a coherent political opposition from forming. Many people with whom I spoke believe that the hiring of Richard Leakey (of the prominent family of anthropologists) to reform the civil service is a hopeful sign that change for the better is finally under way.

Oserian employs 5,500 workers, who support 6,000 dependents. In the context of the limited capacity of government to provide social services, the company has developed a comprehensive labor, environment, and wildlife protection management system to meet the challenges of operating such an extensive enterprise in a developing country.

The first issue Oserian faced was to put together and train a workforce. From field hands, pesticide applicators, sorters, graders, quality control specialists, and security guards to field, greenhouse, and production managers, the operation had to be built from scratch. Base pay for flower sorters is about the Kenyan average of just over $1 per day, but piecework bonuses raise that level to $1.70 to $3.30 per day. Workers are provided with uniforms and, where needed, protective equipment and clothing.

It soon became clear that to obtain healthy, reliable workers, it was necessary to do more than hire, train, and pay them. The company embarked on a comprehensive form of social capitalism. Oserian has built housing for 85% of its workers and dependents on company land that is free to employees. Although simple by Western standards, the apartments are clean, if crowded. The company has provided an electric power plant, a clean drinking water treatment facility, and a sewage disposal system. Also on company land are shops that sell food, small restaurants, and even an employee-run credit union with assets of $1 million that provides low-interest loans to Oserian workers. Transportation on the dirt roads is mostly by the 3,500 bicycles made available at cost by the company. A security force protects the perimeter from intruders (a not insignificant safety consideration) and to maintain internal order.

Because many of the employees are women with children, the company has organized daycare and preschool for 650 young children. Nursing mothers are given two breaks in an eight-hour shift to breastfeed their infants. A full school system through secondary level has been built; teachers were hired by the company. Workers who have been with the company for three years are eligible for full healthcare benefits. An unannounced visit to a company health clinic revealed an excellent primary-care facility that addresses most of the basic needs of the employees, including immunizations, comprehensive maternal and child healthcare programs, tropical disease screening, family planning, and health education. Serious cases are referred to hospitals in Nairobi, and the company transports the sick or injured by air or ambulance. The company also has built a stadium and outdoor sports fields. The official Oserian soccer team is among the top teams in the country.

Although the workers are not organized in labor unions, two levels of worker councils address issues that range from pay to housing to the quality of community services. The councils meet with management every few weeks to discuss issues, and emergency meetings can be convened by the workers or management.

The company faces many environmental, resource, and wildlife issues. Its major reasons for choosing this particular location are year-round sunshine and ready access to water from Lake Naivasha. As noted earlier, the lake is a protected bird migration site that should be subject to special protection under the Ramsar Convention (Susskind 1994). However, the Kenyan government is lax in enforcing environmental standards. The scientific and environmental communities are concerned that the flower farms will overdraw the lake water and then contaminate the lake with fertilizer and pesticide runoff. For the past 12 years, an ecological monitoring and research project at Lake Naivasha has been ongoing; it is headed by investigators from the University of Leicester (United Kingdom) and supported by U.S.-based Earthwatch Institute. The scientists from this project have been reporting their findings and working with flower farm managers to reduce the company's impact on the lake (Harper 1999).

Oserian has installed drip irrigation to reduce its water use (which is remarkable, because 80% of the water is free, unmetered from the lake, whereas the remainder comes from company-drilled bore holes). The company has developed an extensive system of integrated pest management that limits pesticide use to actual threats rather than preventative measures. Similarly, fertilizer use is metered and applied only at specific, predetermined times in the plants' growth cycles. An extensive composting program returns organic waste from the fields to the soil, and 100,000 tons of cattle manure are collected from local tribesmen and applied to the fields each year. Gravel- and sand-lined pits have been dug at the lower edges of fields to trap nutrient- and pesticide-rich runoff so that it cannot reach the lake. Sewage for 12,000 people plus production wastes are treated on site and not returned to the lake (personal communication from Tom Fraser, September 21, 1999,

Naivasha, Kenya). Analysis of the lake water by the independent scientists shows little evidence of contamination from these agricultural activities. Attempts to use natural hot water from thermal springs to heat greenhouses at night and generate electric power were thwarted by the state-owned Kenyan Generation Company, which controls the rights to geothermal resources in the country (unpublished observations from an on-site visit to Oserian Development Company and Earthwatch Lake Naivasha Research Station, September 1999, Naivasha, Kenya).

Because the farm lies between Hell's Gate National Park and the lakeshore, large game animals travel to the lake for water at night. Oserian maintains a wildlife corridor that is more than one mile wide through its property to the lake, where it owns four miles of shoreline. The company also must protect its fields from grazing animals and trampling by large game such as elephants. A nonlethal electric fence appears to have solved the problem. A major tree-planting effort is under way on the property to reforest land that has been cleared for agriculture or wood for charcoal.

Fraser and Oserian's managers made clear that despite the company's elaborate management program, no written EMS was in place. Decision-making is done on a consultative basis among managers and between managers and worker councils as issues arise. Environmental, labor, and social issues are addressed as an integral part of the business operation. Because of the competitiveness of the market, it is essential that operating costs be kept low.

I asked Fraser what drove the company to provide so many social benefits and to go beyond any possible requirements of the Kenyan government. He replied that there were really two reasons. First, to have an effective workforce, worker health needs to be protected and maintained. This objective can best be accomplished by providing housing, access to clean water, sanitary sewers, and a good basic healthcare system. Second, customers in the Netherlands demand that products such as flowers and vegetables be produced under conditions that pay adequate wages, provide safe working conditions, and support environmental quality. The wholesalers and retailers who buy from Oserian demand the right to inspect the firm's operations at any time to make certain that the company is complying with the agreed-upon standards. As a first line of defense, Oserian and most of the other regional growers have created a local inspection team that checks to ensure that standards are being maintained. The inspector who accompanied us believed that one group of workers required protective clothing and noted this on a special reporting form. The management system has been built on the extensive corporate social and environmental program described earlier, with a set of corporate principles and an eye on the competition and the bottom line.

Social Dimensions of EMSs

What are the social implications of EMSs, and what can we learn from the examples that have been examined in this chapter? First, EMSs span a wide range of procedures and practices. They may be either codified corporate systems or sets of practices that operate under an informal set of guidelines. Second, it is possible to create an EMS that focuses only on achieving regulatory compliance in the local physical or natural environment of a particular facility or one that addresses global concerns that exceed current legal requirements. Third, public expectations are growing for companies to address an increasing array of social issues in addition to environmental protection. These additional social values include safety and health; equity, fairness, and justice; community values and participation; security; education; and most important, trust. In the next section, I examine each of these social values in the context of the case studies and try to generalize the implications for the use of EMSs for policy purposes.

Environmental Quality, Safety, and Health

It is important to recognize that environmental quality, safety, and health rank high as social values to the public. It is common to treat these values as technical measures or even technical compliance issues rather than recognize them as being at the forefront of social concern. For any EMS, a positive outcome of these values is the minimal expectation by the public.

All three of the case studies cited rank high in achieving levels of environmental quality, safety, and health of workers and the public. However, the reactions of the public to these efforts are surprisingly different. It can be argued that in absolute terms, the actions by any one of the chemical companies had the most significant impact on environmental quality and on public safety and health. For all of the companies listed, reducing the release of chemicals documented annually by the Toxic Release Inventory, halting the production of CFCs to protect the ozone layer, and addressing climate change have major beneficial impacts on the environment, especially given the scale of operation of these large firms. Also, these chemical firms most closely resemble the formal EMS model in part because they have been members of the Responsible Care program for 10 years or more.

Yet smaller companies have achieved high marks for reducing environmental impact even in the absence of a formal EMS. Stillwater Mining managed to address a full range of biodiversity, air, water, and land protection issues by using a general set of environmental and negotiating principles rather than having a set of specific environmental management procedures in place. The same is true of Oserian, which was hardly concerned with meeting the environmental, water use, or wildlife protection regulations of Kenya. Instead, the company responded to its need for a healthy, productive workforce and to consumer demands that certain values be met. Like Stillwater, Oserian operated on a set of general principles rather than implementing a comprehensive environmental health and safety management plan. This approach may be more characteristic of small start-up firms than of large established companies. Still, Oserian had to comply with criteria to which it had agreed within its own growers organization and the highly specific demands of its customers.

Equity, Justice, and Fairness

To be successful, any system has to be seen as being fair, just, and providing some measure of equity. Equity does not mean equal; it means a fair proportion of the benefits. In the same vein, if environmental risks are present, they should be borne by those who are responsible for the problem or who stand to gain the most by the venture. For example, the outrage over Bhopal was that the people who suffered the most were bystanders, who were poor, whereas the responsible company was half owned by a wealthy American multinational corporation. In the United States, the issue of environmental justice has sparked strong reaction against several firms and local governments that are accused of locating polluting industries and waste disposal in poor less influential communities.

Over the past several years, the alleged exploitation of workers and the use of child labor in developing countries has inspired major consumer outcries. When a pair of running shoes sells for $100 and a person making them in Southeast Asia receives only $1 per day to make several pairs, it is not difficult for the public to become exercised over the inequity of such a division of rewards and demand a change. Clothing and sporting goods companies have been on the receiving end of substantial criticism for unfair, sweatshop, and child labor practices. The most recent example is the U.S. university-led campaign concerning sports uniforms. Nike Corporation responded by transparently identifying its suppliers. Several others are likely to follow, but being the first actor conveys some advantages, such as muting criticism and setting the rules that competitors must follow.

The determination of what is fair and equitable often is viewed through a filter of self-interest. Much of the dismay in the developing world over the issue of climate change is the issue of equity and fairness. The perception is that the rich industrial countries, which have benefited from burning cheap fossil fuels and which are responsible for two-thirds of global emissions, will suffer far less from climate change than will the poor countries and island states that have contributed relatively little to the problem. In other words, the burden is not being equitably shared, and the outcome is unfair. The

United States turns the issue around and says that it is unfair that only industrial countries are being required to reduce greenhouse gas emissions under the Kyoto Protocol climate agreement. The Organization of Petroleum Exporting Countries (OPEC) nations argue that it is unfair to reduce greenhouse gas emissions by lowering the consumption of oil and gas on which their livelihood depends. No justice system seems capable of reconciling these differing perceptions, but it is rather easy to make the case that these three positions do not have equal claims to equity.

As difficult as it is to address, nothing undermines public confidence and support faster than the perception that a process fails to meet equity and fairness criteria. If an EMS-based system is to become the new paradigm, it is essential that it be equitable, just, and fair.

Security

Security concerns of society have multiple meanings and may differ for different groups in different places. The concept of *environmental security* was introduced by Jessica Tuchman Matthews (1989) and has been the subject of much discussion since. At a personal level, individuals wish to feel secure from personal threat, whether it be an attack from another person or exposure to unsafe working conditions or a dangerous chemical.

Clearly, the public demands protection by the corporate community from both catastrophic releases (such as occurred at Bhopal and Minamata) and the pervasive release of chemicals into the environment. It is not yet clear that the chemical industry has learned the lesson of the *Silent Spring*–driven concern over pesticides and other chemicals in the environment. The concern over cancer and nervous system damage engendered by the era of wide-scale release of chlorinated hydrocarbons is being replayed in the debate over the potential threat of endocrine disrupters. The initial evidence of chemical effects on the hormonal system of wildlife is increasing, and the debate over whether humans are threatened has a familiar ring. The public is not likely to feel secure if their hormonal and reproductive systems are under potential assault. Instead of denying the claims made by Colborn and others (1996) in *Our Stolen Future*, the chemical industry instead might respond as it did when the ozone depletion issue threatened its products. At that time, industry organized an open peer-reviewed research process designed to get to the bottom of the issue and to resolve specific unanswered questions. A similar effort to unravel the extent of the endocrine disrupter threat would be highly welcome and would contribute to the public sense of security.

Whereas most companies need not provide the extensive security network that is given to the workers at Oserian in Kenya, the guarded community there provides needed protection in an often lawless society where crime and personal assaults are common. The point is that a company should provide the level of security that is needed for its employees and the surrounding community.

Education

Of the examples we considered, only one firm (Oserian) established a formal education system. In Kenya, a developing country where the public education system has become too expensive for many people, the company makes up for a failure in the public system. By providing a formal school program for the children of its employees, Oserian benefits by making it possible for the parents to work instead of having to care for their children during the day. This level of educational commitment is not needed for all companies to be socially responsible.

Yet all of the companies educated their employees regarding the need to meet environmental health and safety criteria and trained them to do their jobs. What is less explicit but equally important is the ongoing social learning that companies and their employees undergo as they evolve new ways to address problems. Learning by doing is a far more effective way to proceed than to analyze problems to death and then decide that nothing can be done. The immense improvement in corporate environmental health and safety practices during the past 30 years is largely a product of this kind of learning. Experiential learning has been responsible for lowering the costs of achieving compliance and even has produced several economic benefits as inefficiencies that cause pollution have been eliminated. The sustainable development paradigm being pursued by DuPont and other companies is likely to produce the greatest gains in the future. In an EMS-based compliance regime, it will be essential that the regulators not place too heavy a hand on the system, so that it can evolve through an ongoing learning and mutual education process. Without this flexibility, it will be impossible for the continuous improvement required in management practices to evolve in response to new challenges.

Community Values and Participation

The most difficult social aspects for a company to address are aligning its values with those of the community and engaging members of the public, environmental organizations, and other stakeholders in the process of decisionmaking.

Despite the fact that the DuPont Company has moved beyond environmental protection to sustainability principles, its actions have failed to generate the kind of support among the public or within environmental organizations or regulators that its actions might appear to warrant. The reason is complex but appears to lie partly in its failure to engage the public in any meaningful way in its decisionmaking. Most of the company's public announcements have been ad hoc and are seen by the public and environmental organizations as self-serving. A decade ago, when DuPont led the chemical industry into full retreat by abandoning the manufacture of CFCs to protect the ozone layer, critics were quick to point out that DuPont stood to make a larger profit on CFC substitutes because the patent on CFCs had expired. Europeans are particularly critical of DuPont, the U.S. refrigeration industry, and EPA, which they see as raising obstacles to the use of nonproprietary hydrocarbon refrigeration substitutes not only in the United States but also in developing countries.

DuPont's spectacular recent announcement of a 65% reduction in its corporate greenhouse gas emissions has met with criticism that these emissions are from its chemical production facility and although not currently regulated, should be controlled anyway. Also, critics point out that DuPont stands to reap several hundred million dollars in profit by selling emissions reduction credits in a future global emissions trading market. Announcing its action at an industry conference devoted to highlighting the need for companies operating in the United States to receive greenhouse gas emissions credits for taking early reductions only enhanced that impression.

John Elkington points out that the chemical industry has "excellent people who have done extraordinary things, but they do not recognize the scale of the problem they face." The mistake of the chemical industry, he says, is to keep telling the public how the industry benefits them; "people are not persuaded by that" (Reisch 2000). DuPont, like many corporations, has tried to inform the public of its good environmental works through media ads and its annual corporate environmental report. An extensive television campaign consisted of sequences of seals, birds, and other animals clapping as DuPont's clean environment record was recounted to the accompaniment of Beethoven's "Ode to Joy." The environmental organization Friends of the Earth published an extensive rebuttal entitled "Hold the Applause" in which it revealed that DuPont legally released more chemicals into the environment than did many of its competitors. A comparison of six corporate environmental reports found that despite its impressive accomplishments, DuPont's report was short on specifics and consisted of general statements (Whitten 1997).

Unlike DuPont and most large corporations, smaller firms such as Stillwater and Oserian have been much more successful in engaging stakeholders. The ongoing participation of employees, environmental scientists, environmental organizations, and other members of the Lake Naivasha Riparian Association represents a highly participatory process that has produced an array of environmental and social benefits in Kenya. Although such an engaged approach is unusual in the United States, it is much more common in the Netherlands. It is interesting to note that a Dutch firm is translating a Dutch approach to a culture very different from its own.

Trust

In the final analysis, the success of any new system depends on trust. Unfortunately, firms and regulators have often undermined public trust when their actions are seen as in conflict with the public interest.

Monsanto may have lost the current round of introducing genetically modified foods into general commerce and world trade. Even though the company was among the first to establish an environmental advisory board with distinguished, independent thinkers, Monsanto management refused to heed their warnings. European governments and consumers have rejected the importation of genetically modified grains and other foods outright, and opposition within the United States appears to be growing. The company's high-handed approach of intimidating opponents with costly law suits and attempting to suppress hostile journalistic accounts has not won the company much public support. Monsanto's staunch opposition to even labeling products that are genetically modified—and the support of that position by the Secretary of Agriculture Glickman—will undoubtedly go down in history as a classic case of how to alienate the public.

Similarly, energy companies with a sound EMS or ad hoc policies on climate change lose credibility when they continue as members of the Global Climate Coalition. This organization actively works to disrupt the ongoing process of international climate negotiation and blankets American airwaves with misleading advertisements every time an international negotiating session appears in danger of coming to agreement or taking some sort of constructive action. BP-Amoco and Shell have withdrawn from the Global Climate Coalition citing irreconcilable differences with the organization. They remain members of the American Petroleum Institute, which has a much broader agenda than environmental protection but is still active in its opposition to climate change as a public policy issue.

The ACC's Responsible Care program is an early example of an effective EMS that was designed to raise the environmental and safety performance standards of the chemical industry. Unfortunately, the program has lost credibility because the organization has consistently intervened in the legislative and regulatory process in opposition to changes that are supported by some of their member companies. A recent survey in Canada found that only 40% of the public had a favorable view of the chemical industry, and only 18% rated the industry as "excellent" or "good" in honesty (Reisch 2000). Chemical companies have an exceedingly difficult time shedding their past legacy as environmental polluters both individually and as an industry.

Conclusions

The process of improving corporate environmental performance during the past 30 years has emphasized compliance with a set of rules. Good outcomes have been assumed to result if a regulated firm or institution followed the regulations and companies followed prescribed management systems. The United States pioneered public participation requirements, but engagement levels appear to have declined in recent years. Increasingly, the model has become more like the French approach in which technical expertise replaces public input. The public has to some extent transferred its responsibilities to the professionals of the national environmental nongovernmental organizations. In reaction to this trend, local militant grassroots organizations are emerging that are less technically trained and more ideological than the national groups. One need only look at the role of fervent environmentalist and cultural protection groups at the unsuccessful World Trade Organization meeting in Seattle, Washington, in December 1999 to see how unproductive ideologically driven public concern can be.

A major challenge to an EMS-based green tier approach to improving environmental quality will be to engage the public in the process. Companies need to learn how to tap into the concerns and expertise of the communities in which they are located to develop their EMSs. It will be a challenge to have a system that is transparent yet protective of proprietary commercial information. No matter how responsible a company is and how effective its proposals, ex cathedra declarations are unlikely to gain it much credibility and support. A mutual gains negotiation approach among all of the stakeholders will be a prerequisite for success (Susskind and others 1999).

Any EMS used for meeting societal goals will need to have some set of written contracts between the company or industry and the state or federal enforcement agency. Issues of proprietary information need to be addressed but must not be allowed to stand in the way of obtaining such agreements. The negotiated contract system that has evolved over the past decade in the Netherlands provides an excellent model that can be adapted to American culture. Dutch society is as concerned with equity and fairness issues as it is with environmental outcomes and has developed a coherent sense of what equity and fairness mean.

Companies have become increasingly effective at cleaner production. What remains is the need to replace the dirty products that are being so cleanly produced. Environmental problems are increasingly associated with the use of products such as inefficient automobiles and the expanding array of electronic gadgets that have become the hallmark of our era. Product stewardship based on life cycle analysis is one area where corporations can make a major contribution toward improving environmental quality. The accelerated CFC phaseout by the electronic industry is an excellent example of how this might be done. The creation of private green trade regimes and process-based contracts between firms as has been done by Oserian and the retailers it supplies may be the most effective means for expanding best practices to the developing world as well (Moomaw and Unruh 1997).

There will need to be an ongoing learning by doing process. It is not clear just how best to take advantage of the special talents and interests of all of the individual stakeholders who make up the policymaking ecosystem. It is no longer feasible for corporations to exclude these stakeholders from the discussion of how best to meet societal goals.

The most important commodity in this process is trust. It can be built by demonstrating daily that corporations are addressing the social values of the public and the community. Abusing economic and political power will quickly undermine public confidence, especially of large corporations. The greatest threat to the highly desirable green tier approach is the failure to reform campaign financing. As long as the public perceives that companies can buy influence in the regulatory process, they are going to insist on tight regulatory control rather than flexible negotiated compliance systems. This reason may be why a group of Fortune 500 company executives working through the Council on Economic Development recently called for a major overhaul of campaign financing that would limit the corporate political influence that comes with their extensive economic power.

A spokesperson for Shell stated recently that society has expanded its expectations of the corporate role. Companies are no longer expected solely to provide goods or services and show a profit. Increasingly, society expects companies to be innovators and leaders in environmental protection, labor practices, and human rights as well. He was speaking with the voice of experience in response to public outcry over Shell's attempt to sink a used oil rig in the North Sea and its failure to intervene in the repression and execution of protesting Ogoni tribal peoples in Nigeria, where Shell is the major oil producer.

To this observation, I simply add that as we proceed, it is essential to go beyond *environmental* management systems to address the sum of societal concerns through the development of *sustainability* management systems. It is ironic that the extensive gains in environmental quality achieved by the chemical industry have gone largely unappreciated by the public and by most government regulators. A few companies are beginning to link their EMS systems with sustainability goals (Reisch 2000), and if the successes of Stillwater Mining and Oserian Development Company are any indication, this approach can lead to an effective outcome for companies, society, and the environment.

The process we choose to follow will be as important as the environmental outcome. If an effective and transparent process is created for using sustainability management systems, we can look forward to the writing of a new social contract between the private sector and society.

References

Carson, Rachel. 1962. *Silent Spring.* Boston, MA: Houghton Mifflin.

Colborn, Theo, J. Peterson Myers, and Diane Dumonoski. 1996. *Our Stolen Future.* New York: Dutton, Penguin Books.

Harper, David, M. 1999. *Kenya's Wild Heritage.* Expedition Briefing. Maynard, MA: Earthwatch Institute.

Matthews, Jessica Tuchman. 1989. Redefining Security. *Foreign Affairs* 68(2): 162–177.

Meyer, George. 1999. *A Green Tier for Greater Environmental Protection.* June. Madison, WI: Wisconsin Department of Natural Resources. http://www/dnr.state.wi.us/org/caer/cea/reinvention/green_tier/green_tier.htm (accessed Sept. 15, 2000).

Moomaw, W.R., and G.C. Unruh. 1997. Going around the GATT: Private Green Trade Regimes. *Praxis* 13: 67–83.

Reilley, Dennis H. 1999. DuPont Announces New Targets to Reduce Greenhouse Gas Emissions. Press release. September 13. Washington, DC.

Reisch, Mark S. 2000. Responsible Care. *Chemical and Engineering News* 78(36): 21–26.

Susskind, Lawrence E. 1994. *Environmental Diplomacy.* New York: Oxford University Press.

Susskind, Lawrence E., Sarah McKearnan, and Jennifer Thomal-Larmer (eds.). 1999. *The Negotiation Handbook.* Thousand Oaks, CA: Sage Publishing.

Whitten, Patience. 1997. Gap Analysis of Corporate Environmental Reports. Unpublished memo from summer internship. Medford, MA: Tufts University, The Fletcher School of Law and Diplomacy.

7

Information, Environmental Performance, and Environmental Management Systems

Shelley H. Metzenbaum

The way businesses handle their environmental responsibilities has evolved over time. Some companies have become more aware of and accountable for the environmental consequences of their actions. As they have done so, they have begun to recognize the value of taking a more systematic and, in a few cases, even a strategic approach to environmental issues[1] (Hoffman 1999). To support the transition to systematic and strategic management of their environmental responsibilities, many companies and government facility managers have begun to adopt *environmental management systems* (EMSs), defined as suggested by Nash and Ehrenfeld in Chapter 3: "formal structures of rules and resources that managers adopt to establish organizational routines that help achieve corporate environmental goals."

To take advantage of this shift, some governments have begun to experiment with ways to recognize EMS use in their environmental regulatory schemes. In 1993, the European Union established its Eco-Management and Audit Scheme (EMAS). EMAS encourages companies to adopt voluntarily a policy committing to "continuous improvement in environmental performance" and a site-specific, independently audited EMS to support that policy (Morrison and others 2000). The European Union also requires member countries to promote EMAS. Furthermore, it expects them to establish an organizational body and system to accredit independent verifiers who can confirm the compliance of EMAS-registered sites (Gouldson and Murphy 1998). The United Kingdom was the first government to adopt an EMS standard, in 1994.[2] As of June 2000, six states and the U.S. Environmental Protection Agency (EPA) had begun to offer incentives to companies that adopt EMSs or their equivalent and fulfill other "leadership" requirements that vary by state. Another six states are considering similar programs to reward companies that take certain "leadership" actions that include the adoption of an EMS (Crow 2000; U.S. EPA 2000).

In 1997, the International Organization for Standardization (ISO), which has long played a central role in establishing standards in support of total quality management systems, established ISO 14001 as its first environmental management standard (Gouldson and Murphy 1998, 60). The importance of the ISO environmental standard was significantly elevated in 1999 when Ford and General Motors announced that they would require their suppliers to be certified (or registered) to the ISO standard (ENDS Report 1999).

These actions raise an important question: does the use of an EMS improve environmental performance? That is, do companies that use an EMS have a less negative impact on the environment (or perhaps even a positive one) than companies that do not? Or do they at least have a better record of compliance with environmental laws?

An affirmative answer would help government decide whether to reward companies adopting EMSs—or even require EMS adoption—to improve public health and environmental quality. It would help a business decide whether to adopt an EMS to manage its environmental responsibilities. It also would help citizens assess whether they want neighboring companies to adopt an EMS and whether to have greater confidence in those that do.

Answering this fundamental question—whether EMS use leads to environmental gain—requires attention to a few more. Most basically, what information is needed to determine whether use of an EMS enhances environmental quality? If an EMS can indeed improve environmental quality, what information is needed to support government, business, or citizen use of EMSs as environmental protection instruments? Furthermore, is the existing information infrastructure—the processes and players that generate, transform, and deliver that information—likely to produce and deliver information about EMSs to those who need it, when they need it, in a form they can use?

One additional policy question remains. If EMS use does not predictably lead to environmental gain, will widespread use of EMSs at least generate more credible and comparable information about environmental impacts that will strengthen the existing environmental protection system? Over the past decade, numerous environmental, business, government, and community leaders have come together in multiple settings to explore ways to improve the environment and the existing environmental protection system. A common theme emanating from these groups is that the existing system needs to be more performance-focused, information-driven, flexible in the means of meeting standards, strictly accountable, and open and transparent

(NAPA 1995; NAPA 1997; E4E 1998). Will increased use of EMSs generate the underlying environmental performance information essential to performance-focused, information-driven environmental protection strategies, allowing increased learning about the effectiveness of various intervention strategies, public performance comparison, performance agreements, or increased use of performance minimums?

Before policymakers rush to embrace EMSs, these questions need to be addressed, particularly about whether use of an EMS predictably improves environmental performance. Surprisingly, we are far from a clear answer to this question. Initial research on one kind of EMS suggests that its use does *not* ensure environmental gain. Even more surprisingly, EMS use also fails to ensure progress toward a performance-focused, information-driven system likely to drive environmental gain.

These findings may apply only to the particular EMS studied. However, they underscore the need for more experience with the use of EMSs—and more information and research on the environmental and systems impact of that experience—before the government embraces the EMS as a policy tool. To ensure that reliable and credible research on EMSs and their impact can be generated, far more attention needs to be directed to the environmental information infrastructure. A robust performance-focused information infrastructure is prerequisite for government and citizens to be able to assess the benefit of rewarding or mandating the use of EMSs.

The strength of the information infrastructure is not an issue for government alone. Businesses and other regulated entities interested in using EMSs also need to know what kinds work and under what conditions. They, too, need to invest time and attention in building an infrastructure that will generate credible information, especially if they want to use EMSs to improve their relationships and reputation with the public.

An information infrastructure that generates credible and comparable information about environmental performance not only will help answer questions about the value of EMSs as a policy or business environmental protection tool. It also will help build the foundation for a more effective and transparent regulatory system, one focused on improving environmental results while allowing greater flexibility in the means to achieve them, greater fairness in holding individual pollution sources accountable, and greater public involvement in and awareness of decisions that affect them. A solid information infrastructure is essential to an "information-driven, performance-focused" environmental protection system.

In the remainder of this chapter, I explore the questions just presented. First, I consider what a performance-focused, information-driven environmental protection system is, its expected benefits, and why attention to the information infrastructure is important. Next, I turn my attention to what we currently know about the benefits of using an EMS, what we want to know, and whether the existing information infrastructure lets us answer those questions. Finally, I suggest what the essential elements of a robust performance-focused environmental information infrastructure are and briefly explore possibilities—including the value of increased use of EMSs—for strengthening those elements.

What Is a Performance-Focused System, and Why Is It Important?

Before considering whether EMSs will improve environmental quality or at least contribute to a performance-focused, information-driven system for environmental protection, it is useful to clarify what this system is and the expected benefits associated with creating such a system. As the name implies, a performance-focused, information-driven environmental protection strategy embraces two distinct concepts: a focus on performance and a heavy reliance on information.

Focus on Real-World Outcomes

Focusing on performance does not mean the same thing to every person. To many, a performance-focused system is simply one that sets goals and monitors progress toward those goals. It does not matter what type of goal it is; the goal can be an activity target (for example, number of phone calls answered) as easily as it can be an outcome target (for example, safety levels on the highway). Thus, many people would consider a system that sets sales goals for a sales force or enforcement action goals for environmental regulation a performance-focused system.

A goal-focused system is not the notion that advocates of performance-focused environmental protection strategies have in mind. They believe goals should be framed and measured in terms of outcomes, primarily environmental and health outcomes and impacts. (Most public organizations also seek to advance fairness and a healthy democracy as a complement to advancing their mission-specific goals.) As stated in one highly respected report, an improved system would "foster the creation of environmental goals and milestones, and, where appropriate, use performance-based requirements to achieve them.... It would tolerate no rollback in protecting the environment and human health ..." (Ruckelshaus 1998). It implies defining goals in terms of specific improvements to environmental and public health quality, reductions in releases to the environment, or lessened environmental or public health risk.

Advocates of performance-focused systems believe that just setting goals and monitoring progress toward environmental and health outcome targets is not enough. Goals and progress toward the goals should guide—and drive—decisions and activities. Under a performance-focused permit, for example, a business would commit to operating within an emissions or discharge limit rather than to installing a specific control technology (U.S. EPA 1999, 8–9). Similarly, a performance-focused state agency would send inspectors to check out sources believed to be big polluters (or that have poor compliance records) rather than facilities with historically strong compliance and environmental records. The U.S. Environmental Protection Agency (EPA), adopting a performance-focused strategy, might focus on ensuring that water quality is adequately monitored before reviewing state or community watershed plans.

Under a performance-focused system, process and activity targets (for example, number of inspections) can be useful, but only when a plausible link has been established between the activity and the outcome goal. For example, an outcome-focused activity target might call for inspecting all sources or monitoring key points in a severely polluted watershed to improve its water quality by 10%. Activity targets help only when a performance-focused system continually loops back to ensure that assumptions about the link between the process or activity targets and the anticipated environmental and health consequences are valid, and updates those assumptions as experience suggests is appropriate.

That is not to suggest that business and government would want to ignore activity levels and processes altogether. Environmental agencies might want to track activities such as the completion of watershed plans, inspections, and enforcement levels. They could then use their information to analyze why different outcome levels occurred. They would not, however, focus on activity levels as targets in their own right.

The value of focusing on performance is neither a new nor an astounding proposition. Indeed, many federal environmental laws embrace the notion of environmental outcome and impact targets. The Clean Water Act established as a "national goal that the discharge of pollutants into the navigable waters be eliminated by 1985." The Clean Air Act requires EPA to establish minimum air quality goals that states must ensure their communities meet. The Resource Conservation and Recovery Act sets a performance target of zero migration of hazardous waste from a disposal unit or injection zone. But the performance emphasis has gotten lost in the past 30 years of implementation. The reality is that few government agencies and regulated parties have had the interest, political support, and financial wherewithal to measure environmental and health improvements while meeting statutory (primarily process) deadlines established by the U.S.

Congress, state legislatures, and—for states, localities, and the regulated community—EPA.

Phenomenal advances in technology over the past few decades have made measurement, analysis, and communication of progress toward outcome goals more affordable than when many of the process-focused policies were established. However, updating those policies takes time and attention. Few regulatory agencies have the time or money even to keep up with new legislative mandates and new scientific findings, so it is not surprising that process-focused policies are seldom updated. However, the advances have been so dramatic that it is timely to recapture the value of environmental and health outcomes as the primary decision driver.

Performance-focused approaches tend to drive improvements in environmental outcomes in several distinct and complementary ways. One reason almost seems too obvious to mention. If an organization does not know what its goals are for environmental and health outcomes and impacts, and cannot measure whether it has achieved those goals, the organization cannot possibly determine whether it is making progress and whether its intervention strategies are working. Goals clarify direction. They motivate and hone organizational focus. Outcome-focused goals encourage innovation and adaptation. Goals also aid communication, helping to enlist allies to advance toward a goal. Publicly announced goals introduce even greater benefits, because external observation can be a powerful motivator. So, too, can a carefully structured reward system.

Goals Motivate

Social psychology research on the motivation of individual workers finds that those given specific and difficult goals outperform those given a "do your best" goal or no goal at all.[3] There are several reasons for this. "Goals provide us with a clear direction; inform us that we need to try hard; remind us that an end is in sight; and encourage us to think about the process of reaching that end" (Katz 2000, 1). Thus, the simple act of establishing a goal (outcome or otherwise), even without linking it to rewards, tends to motivate improved progress toward that goal.

Organizational Focus

It seems reasonable to presume the motivational effect of goals on individuals will translate to the organizations for which they work. The exception would be when individuals within an organization work toward conflicting or unrelated goals. Under such circumstances, heightened motivation might not result in increased performance. Clearly stated organizational goals address this problem by increasing the chance that workers and work units within an organization are aware of organizational goals, will focus and work on them rather than their own goals, and will make progress toward them (Kaplan and Norton 2001). Essentially, goals play a communication and attention-focusing function that is especially valuable for larger organizations.

Innovation and Adaptation

Outcome-focused goals not only motivate progress toward an outcome goal; they also offer flexibility, which allows innovation and adaptation. Outcome goals, rather than process or activity goals, allow workers and organizational units the flexibility to determine how to meet those goals. This flexibility makes innovation and case-specific adaptation easier and faster, because it does not require external review and approval. For example, when the Clean Air Act placed emissions caps on sulfur dioxide at half their historic levels, it spurred the railroads to develop new technologies that reduced the cost of delivering low-sulfur fuel to utilities (Swift 1997). Innovation also brings productivity gains, making it possible to attain a goal at a lower cost or achieve even greater outcomes for the same cost.

Engagement and Enlistment for Shared Goals

Disclosing goals to the public further enhances the likelihood of achieving improved outcomes. Publicizing goals can engage and enlist those beyond the boundary of a single organization who share the organization's goals. Focusing on shared goals enables multiple parties to assess skills and resources needed in pursuit of a shared goal, and agree on who can and will contribute what by when. This strategy has long been used in the Chesapeake Bay and, more recently, has proven remarkably effective on the lower Charles River in the Boston area. When the EPA regional administrator announced that the Charles River would be fishable and swimmable by 2005 and supported that goal with the management resources to support and coordinate a multiparty effort, it enabled the agency to enlist multiple cooperating partners and to encourage cooperation from others, resulting in dramatic increases in water quality in a relatively short time (Metzenbaum forthcoming).

Motivation from External Expectations

Beyond engaging actors outside organizational boundaries, revealing goals publicly can also heighten internal motivation and lead to improved outcomes. Even the most self-motivated individuals find it useful to enlist external observation and make external commitments to boost motivation. Weight Watchers, Jenny Craig, and other diet-support companies are an entrepreneurial testimony to people's recognition that external disclosure of performance goals and progress toward them is a powerful motivator.

Savvy managers understand they can fruitfully use external attention as a way to encourage subordinates to pursue goals seriously. For example, when the secretary of Florida's Department of Environmental Protection introduced the agency's new quarterly performance report, she attached to the first report a cover memorandum addressed to her division chiefs. The memorandum not only commended each division for its specific accomplishments but also identified priority items needing work. Each division chief was also instructed to "prepare a one page course of action ... [for] each of the focus areas noted" to be shared with the governor and the press (Wetherill 1997). External observation essentially introduces a subtle incentive into the system—the pride or embarrassment that ensues when a publicly announced goal is or is not attained.

Incentives

Goals linked to more explicit incentives can further motivate workers and organizations to improve their performance. The Clean Air Act Amendments of 1990 link rewards (grants) and sanctions (threatened withholding of grants and limits on local construction projects) to achievement of certain environmental outcome goals. These incentives motivated more aggressive governmental action to improve the environment than did the earlier version of the law that set goals without consequences for nonattainment. The goal-focused law was effective, but not effective enough. Incentives have added valuable additional motivation.

However, caution needs to be exercised in introducing incentives. If accurate performance measurement is difficult or the incentive system inappropriately structured, unintended consequences are likely to arise. Poorly structured incentive systems can cause workers and organizational units to pursue dysfunctional behaviors. They might, for example, carry out more of the activities that can be measured easily and fewer of those that cannot, even though both activities need to be done to improve outcomes. They also might try to manipulate the goal-tracking system (Austin 1996).

In sum, outcome-focused performance goals are likely to improve public health and environmental quality by focusing an organization, allowing flexibility, encouraging innovation, motivating individuals, and engaging the assistance of actors beyond the organization's boundaries who share similar objectives. Performance goals can be especially motivational when the goals are revealed to external observers. Incentives can further motivate improved performance, but caution must be exercised in structuring incentives so as not to induce dysfunctional behavior.

The question for policymakers considering policies to encourage the use of EMSs is whether increased use of EMSs helps build a performance-focused system that is likely to motivate progress toward improved environmental and health outcomes in the ways described earlier. This question is considered later.

Information-Driven Strategies

What are information-driven strategies, how do they relate to performance-focused strategies, and how do they improve environmental quality? As with the term *performance-focused*, the idea of an information-driven system means different things to different people. Most seem to agree that it implies an information-rich system that generates a broad range of information to inform decisions that affect the environment and public health.

Both performance-focused and information-driven strategies need information about environmental and health outcomes. A performance-focused system requires feedback about the outcomes for which goals have been set to enable individuals and organizations to adjust and calibrate their activities to influence outcomes. Without performance information, they would have to operate in ignorance, possibly running full-speed ahead in the wrong direction!

Whereas a performance-focused system requires information, an information-driven strategy does not require explicit goals to motivate performance improvement. Information by itself, even without the establishment and monitoring of goals and even without any links to incentives, can lead to environmental gain. It can drive improvements in public value in several distinct ways. It can motivate improvement through feedback to the performer. It can inform selection among multiple options to meet the needs and desires of those making the selections. And it can guide political action in a democracy.

The potential of information to drive performance gains has long been recognized. Reliable information is the cornerstone of a vibrant market economy and the keystone of a healthy democracy. Savvy information analysis and management propelled the growth of the modern corporation (Chandler 1977). Federal laws mandating open, credible, and comparable business financial information (standardized public reporting, third-party verification, and penalties for misrepresentation) have long fueled America's thriving capital markets.

Improvement through Feedback

Providing individuals with feedback about their performance levels motivates them (Katz 2000). Therefore, providing environmental performance information—by itself and even without setting a goal—can lead to environmental gain. Feedback not only inspires workers to work harder but also enables them to work smarter—adjusting and calibrating their actions based on information about prior experience.

Feedback can be especially effective when it is about real-world outcomes and people are intrinsically or altruistically motivated, as are many people who work in the environmental field. In these circumstances, workplace accomplishments also advance personal goals. For example, feedback that lets environmentally committed workers know that their efforts have improved the safety of the local water supply is likely to be far more affirming than feedback about the number of plans they have reviewed.

Organizational Learning

In an organization, it often is necessary to make work adjustments in concert with others. Feedback can lead to improved organizational performance if workers share information and plan work adjustments collectively. In larger organizations, the task of assessing what works and what doesn't—and developing, testing, and evaluating new strategies—necessitates systems for collecting, organizing, and analyzing outcome-relevant information from across the organization. If the information gets distributed to people who can and will do something about it, then environmental feedback information can improve the organization's environmental performance.

Comparison

When feedback is comparative, it can be very powerful. It allows workers or organizational units to benchmark with others and enables them to emulate the strongest performers. Comparative performance measurement also acquires the motivational value of goal-setting, because the leader's performance functions as a constantly updating target for everyone else. This can be true even when comparative information is privately disclosed (for example, consultant reports showing how a company or organization fares relative to industry leaders) because personal pride (and in the private sector, the threat of competition) is a powerful motivator.

In using comparative performance information, caution must be exercised not to discourage collaboration. If rewards are given only to individual high performers, for example, low performers may withhold information from them. If rewards are given to high-performing groups instead, it may lessen the motivation of individual high performers within the group. Or, high-performing individuals may pressure low-performers so intensely that it demotivates them, bringing down overall performance rather than raising it up (Katz 2000). For this reason, it is often preferable—especially where altruistic motivation is likely to be high—to make comparative information available without linking it to incentives. Those being measured can then work with their colleagues to improve outcomes. Because they do not feel individually threatened by student test scores, for example, fifth-grade teachers in one Tennessee school redesigned the way they teach science to allow teachers whose students tested highest on science exams to teach science to all fifth-grade students (Gormley and Weimar 1999, 158).

Public Disclosure Drives Environmental Gain

Public disclosure of comparative performance information can have the same effect as public disclosure of goals and progress reports. The pride of being at the head of the pack and the embarrassment of being at the tail can add to the motivational effect of comparative information. In the environmental field, the public disclosure provisions of the federal Toxics Release Inventory (TRI) demonstrate this effect. Numerous business leaders have shared anecdotes about how they took action to reduce emissions when their companies ranked as high emitters on local TRI lists.

Comparison is not always necessary for the power of non-goal-related performance information to be realized. Simply disclosing to the public the environmental impact of private actions can have a significant motivating effect. This is evident in the way business has responded to the disclosure requirements of California's Proposition 65.[4] Proposition 65 requires companies to provide a clear and reasonable warning before knowingly and intentionally exposing anyone to a listed chemical. The Environmental Defense Fund (EDF 2000) found significantly greater reductions in emissions levels in California than in the rest of the nation for the same chemicals since Proposition 65 has been in effect. Public information about environmental impacts, even without requiring goal-setting, can drive emissions reductions.

Open information has served as an important check on the abuse of public and private power as well as the abdication of government and private-sector responsibilities. The notion that business and government should operate in the open is fundamental to the American political and economic system. Justice Louis D. Brandeis wrote, "publicity [is the] remedy for social and industrial diseases. Sunlight is said to be the best of disinfectants; electric light the most efficient policeman" (Brandeis 1914). Federal, state, and local laws have supported the collection and dissemination of information to advance the public interest since the mandate in the U.S. Constitution to count the population. Many laws also ensure citizens access to information about business and government behavior to inform political choice and, where warranted, enable citizen suits to enhance public and private accountability and protect against malfeasance.

Context

Comparison also helps information interpretation. Without context, measures lack meaning. Most measures take on meaning because of their relative value. A runner's time means little to those unfamiliar with average or winning times for that particular race. Learning that a contaminant constitutes x parts per million is an incomprehensible fact for many people. It needs to be put in context—compared, for example, with the y parts per million concentration that EPA or the state considers safe, or at least to the parts per million emitted by other companies in the same industry. When comparisons with similar situations are not possible, measurement of progress toward performance targets provides some context for assessment. Information needs context to aid the interpretation of its importance and relevance.

Informing Selection among Multiple Options

Non-goal-related performance information is also helpful to those who must choose from among multiple options—whether purchasing goods, services, or investments; formulating implementation strategies or policy positions; or voting for political candidates. It helps people and organizations make smarter choices and guards against private parties taking actions that might be detrimental to social outcomes.

Information about a product's environmental impact, for example, helps people who care strongly about the environment find the best match for their personal preferences. The increasingly widespread mention of the EPA Energy Star labels in advertisements for appliances and computers suggests that for some companies, environmentally sensitive buyers are numerous enough to warrant targeted marketing. Generally speaking, providing information that makes it more likely for people to get what they want adds value to society. When some people want products that are more protective of health and the environment, the availability of information about the environmental impact of a product allows the purchase of more environmentally benign products, leading to better environmental outcomes. (Of course, if buyers want products because they harm the environment, the availability of environmental impact information might have the opposite effect; I assume such buyers are few.)

Information about the environmental performance of companies helps environmentally sensitive investors screen out companies with poor environmental track records (Social Investment Forum 1997). It also informs investors who believe that past environmental performance is a useful predictor of future financial performance (Repetto and Austin 2000). Investment directed to firms with stronger past environmental performance ultimately makes more capital available to those firms than to firms with weaker performance. This strategy should, in turn, lead to stronger environmental outcomes.

Environmental information also allows companies and government regulators to assess the relative effectiveness (performance impact) of different intervention strategies for different kinds of organizations. Just as retrospective analyses of business strategies allow businesses to improve their financial performance, retrospective analyses of environmental protection strategies can help identify strategies that are relatively more successful and cost-effective than others. It also can help government look for patterns and assess the prevalence and relative importance of problems that merit priority attention (Sparrow 1994).

Triggering Political Action in a Democracy

Sometimes, public disclosure of comparative information reveals unacceptable levels of environmental performance by a company or a government but no alternate provider is available. For example, the citizens of a state only have one state government, and there is only one EPA at the federal level. If either is performing poorly, the best recourse may be some form of political or legal pressure. Similarly, when a neighborhood learns a local company is dumping waste into a nearby stream, it may have no means of persuasion other than political or legal protest. In such cases, information triggers nonchoice forms of persuasion, such as political protest or legal actions, to compel the correction of unacceptable environmental performance.

Fairness

Credible and comparable public information about environmental performance also can enhance the fairness of the environmental protection system. Under the current system, government regulators make decisions based on the relatively sparse information available to them about which facilities to inspect, which records to review, and how to respond to their findings. Increased availability of public environmental performance information will allow government to allocate its attention more accurately and fairly to firms that pose the most significant environmental problems, with the greatest potential for environmental gain, or that are most likely to be out of compliance. It also will enable companies and citizens to assess whether some firms get a disproportionate amount of attention relative to their environmental impact or compared with other similar firms, and offer analysis and advice to government regulators for redirecting their activities more fairly.

Environmental Gain

In sum, performance goals and measures can lead to environmental gain in several ways. They can motivate improvement through feedback, learning, comparison, public pressure, and incentives. They can strengthen organizational focus and the ability of an organization to enlist the support of others to enhance environmental outcomes. They can inform the choices of people buying products, investing capital, or choosing implementation strategies, increasing environmental value to decisionmakers as well as the broader population. Finally, they can trigger political action in a democracy, allowing people concerned about health and environmental issues to press for environmental gains, and enhance the fairness of the existing system.

EMSs, Performance, and Information Availability

The preceding discussion suggests that EMSs that incorporate targets for improved environmental outcomes, publicly disclose their environmental performance information, and use credible, comparable performance metrics could have a strong positive effect on environmental quality.

Even if the existing evidence does not find that use of an EMS is likely to improve environmental outcomes, government might want to encourage use of EMSs that focus on real-world environmental outcomes and generate greater quantities of credible and comparable environmental performance information. Such EMSs would generate information helpful to managers, workers, consumers, investors, and the public. It also would allow government to establish programs to reward strong performers or negotiate performance agreements to reap greater environmental performance in return for greater flexibility.

Why Is an Information Infrastructure Important?

Despite its enormous potential, the full value of information often is not fully realized because getting information to would-be users in a timely and attention-getting manner can be a difficult and costly undertaking. Specifically, some provider must generate the raw information—a story, case study, or some form of measurement—that the information audience needs and can trust. Then, someone must transform that information into a format the information audience can use, filtering out irrelevant and excessive details so that users will perceive and focus on what is relevant. The information must be analyzed, translated, and packaged so that the intended audience will notice and accurately interpret it. Some provider must disseminate information to users through conferences, publications, the Internet, or whatever medium works. Finally, the intended audience must pay attention to it and use it. Together, these steps comprise the *information infrastructure.*

When the information infrastructure works well, it can be transformative. The U.S. agricultural research and extension system demonstrates the potential of a well-functioning information infrastructure. By creating agricultural experiment stations, extension services, and land grant universities throughout the country (Evenson and others 1979), the federal government put in place the backbone of an integrated information infrastructure that provided farmers and their suppliers with reliable information about best agricultural practices for more than a century. The government-catalyzed information production system was significantly supplemented and complemented over the years not only by private information-producing firms with for-profit interests (such as seed suppliers and the trade press) but also by community institutions (such as fairs and farm clubs). The resulting information infrastructure successfully produced, transformed, and disseminated information that dramatically increased the productivity of the American farmer and dramatically reduced the cost of food for the American consumer.

In the environmental arena, the TRI is frequently cited as the poster child of an effective information infrastructure.[5] Under the TRI, the federal government mandates and collects standardized information about toxic releases of facilities and then makes it publicly available in a user-friendly format. The way EPA managed TRI information represented a significant change in EPA's information infrastructure. Previously, the agency collected and stored all of its business and ambient environment data in large mainframe databases accessible to very few people. When the TRI was implemented, advances in computer technology enabled EPA to make data available to citizens for use on personal computers. Providing the TRI on a diskette allowed users to customize their data analysis to meet their own needs. Newspapers and local public interest groups quickly scrambled to report the names of local companies leading the list of toxics releasers, and many companies, in turn, scrambled to reduce their releases of toxics so they could move down or off the list.

In contrast, the absence of an information infrastructure—particularly the will, capacity, and interest to generate, transform, and disseminate information to users in a timely and comprehensible manner—can preclude the realization of enormous private and public benefits.

In many cases, initial attention needs to be directed toward strengthening communities of information owners and users: "to leverage knowledge we need to focus on the community that owns it and the people who use it, not the knowledge itself" (McDermott 1999). The existence of a driver or multiple drivers with not only the resources and ability but also the interest and motivation to generate the needed information is essential. Too often, such drivers are missing in action. For example, the existing information infrastructure for health maintenance organizations (HMOs) is unable to deliver information to HMO customers in the way they need it, impeding the healthcare market. In the 1970s, as part of an industry effort to stave off federal reporting requirements, the HMO industry created the National Committee for Quality Assurance (NCQA) as an independent nonprofit organization to gather quality assurance information about HMOs. Unfortunately, NCQA lacks the power of a government authority or a near-monopoly customer to compel HMO reporting. Thus, it receives limited information that is of limited value to HMO users (Gormley and Weimar 1999, 47–54). In recent years, frustrated by the weakness of the NCQA system, employers in a few communities have begun to step forward to sponsor and compel HMOs to report comparative data on their performance (Pham 1999). However, this fix only works in the few locales where employers have sufficient market clout to ensure full and honest reporting.

Information infrastructure gaps arise when information needs cross organizational boundaries, creating too many drivers going in different directions without paying attention to each other. Professor Peter Senge of the Massachusetts Institute of Technology has demonstrated the difficulty of diagnosing and fixing problems when multiple parties are dependent on each other and fail to share information across organizational lines. Using a simulation exercise that involves a beer manufacturer, a retailer, and a wholesaler, Senge (1990, 26–54) demonstrates how the retailer's decision to increase its beer order eventually leads to a serious oversupply of beer for all three parties, because each person makes decisions based solely on the information within his or her own sphere of operation. If instead the wholesaler asks the retailer why it is increasing its order or the retailer is able to understand the situation the beer producer sees, the overstock situation can be avoided.

An effective information infrastructure often needs to penetrate organizational boundaries, even when no single entity exists with the authority to control all the organizations. A driver must come forth or be created that can assume the authority needed to gather information from multiple organizations. This driver could be one of the parties involved. It also could be an information entrepreneur who sees the potential profitability of selling boundary-spanning information to multiple organizations or an impartial convener who can help all parties look across their organizational divisions. When no private entity steps forward to fill the gap, it often is appropriate and necessary for government to play the role of convener or to stimulate information entrepreneurs through activities such as information standardization.

Information infrastructure gaps also can arise when the needs of the information suppliers are not sufficiently considered. Many EPA databases have been severely weakened because of this problem. States, suppliers of much of the data in EPA's databases, do not believe that they get much value from the data they supply. EPA generates relatively few reports using the data to help the states (or EPA or the public) understand the quality of the environment, the quality of data about the environment, or how to use government environmental protection resources more effectively. EPA has tried both coercion and persuasion to encourage states to improve their data quality but has been unable to motivate many states to supply more than the minimum data required because the states get so little value from their data entry efforts.[6] To lessen this gap, EPA needs to consider how it can offer states a greater return on their investment by analyzing, packaging, and returning the submitted information in a way that provides greater value to the information-supplying states.

In sum, information can add enormous value in both private and public spheres, but only if sufficient pieces of the information infrastructure are in place to ensure delivery of the right information to the right place at the right time. An effective information infrastructure requires not only an understanding of the key processes in the information delivery system (generation, transformation, and delivery); it also requires that organizations exist with the interest, capacity, and motivation to meet the needs of the potential information audience. Those organizations need not be, nor are they likely to be, a single entity. Indeed, different organizations—from the public and private, for-profit and not-for-profit sectors—can and hopefully will respond to the needs of different audiences. However, if they do not, government should step in to ensure that the gaps are filled, and industry and nonprofit organizations should wholeheartedly support this government activity.

Will Increased Use of EMSs Benefit Government (and the Public It Serves)?

The policy question that needs to be examined is whether increased use of EMSs is likely to improve the quality of the environment or whether it will, at least, strengthen the information infrastructure that can answer that question and support a performance-focused, information-driven environmental protection system.

To answer this question, consider first what government regulators want to know about the value of using EMSs as a tool for improving environmental performance. The fundamental question that concerns environmental policymakers is whether the use of an EMS improves the effect of business on public health and the environment. Beyond mission effectiveness (in this case, public health and environmental gain), U.S. policymakers also often seek policies that advance other societal goals, which include a healthy democracy and fairness. If the use of EMSs promotes these goals, it also would be of interest to policymakers. If not, will increased use of EMSs at least advance the development of a performance-focused, information-driven environmental protection system? If the answer to any of these questions is yes, then government might want to mandate the use of EMSs as a policy tool or encourage it by offering incentives for EMS adoption.

Two recent studies on the environmental effect of using an EMS allow a preliminary answer to these questions. Both studies conclude that the use of an EMS does not predictably lead to better environmental outcomes. They both examine one kind of EMS, the Responsible Care[7] program promoted by the Chemical Manufacturers Association (CMA; now the American Chemistry Council), but use very different research approaches.

In one study, King and Lenox (2000) look at 10 years of data and compare changes in toxics emission levels at companies that participated in Responsible Care with changes at companies that did not.[8] The results suggest that whereas the level of toxics released declined significantly for the industry as a whole, the rate of environmental performance improvement for Responsible Care subscribers did not, on average, exceed that of other chemical manufacturers.

The second study (Howard and others 2000) takes a very different approach, conducting interviews of managers at 16 firms that implemented Responsible Care to ascertain the level and nature of changes at each firm attributable to the program. The authors focus on operational behavior changes based on the presumption that certain behavioral changes are a necessary precursor to outcome changes. They find that the Responsible Care program dramatically changed the way of thinking in three of the firms and was a useful and important environmental health and safety management tool in another three. However, in 10 of the firms, Responsible Care primarily helped relationships with external constituencies without significantly changing internal behavior. In sum, the study finds that the use of EMSs changed operational behavior linked to the environment in some firms, but not in most. In other words, any policy that rewards the use of EMSs without restriction is likely to have significant problems with free riders.

Perhaps our work is done. We had a policy question: does the use of an EMS improve the effect of a business on public health and the environment? We have a policy answer: not for most facilities. These two studies of a valuable public policy experiment undertaken by CMA suggest that it is premature to adopt a policy mechanism that rewards or mandates the use of EMSs.

It is worth sounding one additional note about the use of EMSs as a government policy tool. At one time, a firm's use of an EMS may have been a good indicator or proxy of its environmental intentions, signaling a strong environmental commitment. However, the possible signaling value of EMS adoption was lost when corporations and government started adopting policies that rewarded or mandated EMS use.

Should we close the door on this policy option altogether, or would that be premature? Howard and others (2000) showed positive change in 6 of 16 firms studied. These findings suggest that the way the Responsible Care program was implemented in a particular company might affect the environmental value of using an EMS. Moreover, it may be inappropriate to extrapolate findings about the effect of a chemical manufacturer's use of a particular EMS to other industries, other forms of EMSs, or environmental outcomes other than the level of toxic emissions. Additional experience and research are needed to shed light on whether particular characteristics of EMSs—or characteristics of adopting companies or industries—are likely to affect a range of environmental outcomes (toxic releases and other aspects of environmental performance).

What kinds of characteristics might affect how an EMS translates to the environmental bottom line? One is the information on which it focuses. The two studies cited above look at the Responsible Care program, which calls for participating companies to measure inputs, not environmental outcomes. (CMA updated the Responsible Care program in 1999 to make it more performance-focused, requiring companies to establish and publicly report on progress toward performance goals.) An EMS that focuses on environmental outcomes is likely to have a greater effect on improving those outcomes than one that just tracks inputs. Unfortunately, even the dominant EMS standard, ISO 14001, does not require a focus on environmental outcomes. As noted in a recent study on ISO 14001 (Morrison and others 2000, 47),

> a registered ISO 14001 environmental management system assures that the organization has the capacity to meet whatever environmental performance levels it has set.... It should also be noted that specifications within the ISO 14001 standard do not guarantee that 14001-certified organizations will actually improve their environmental performance; the standard only requires improvement of the management system itself, guided by the assumption that process improvements will lead to substantive improvements. It is conceivable that improvements in environmental performance could be negligible even after full implementation of an EMS....

Consider the possible dangers that could arise if an EMS failed to make improved environmental outcomes a primary objective of the system. An EMS focused on reducing environmental costs without some sort of counterbalancing emphasis to encourage environmental gain risks motivating operating units to cut corners in environmental protection. Similarly, a system focused exclusively on risk might encourage businesses to sell their riskier enterprises, removing the risk from the books of the business but not from the public balance sheet.

How and where an EMS is implemented within an organization also may influence a firm's environmental performance. Again, various characteristics may be important: the level of senior management commitment, whether individual operating or centralized environmental health and safety units are expected to implement an EMS, whether environmental costs are charged to overhead or assigned to each unit, how organizational units are held accountable for their environmental performance, and whether environmental performance information is public.

The way EMS-generated information is broken down and grouped is likely to be another relevant determinant of effectiveness. A case study of the S.C. Johnson Wax Company shows how looking at environmental costs by product line and further disaggregating product lines to distinguish among different chemical classes led to smarter management decisions than would have occurred without such a refined analysis (Ditz and others 1995, 139–156).

The stringency of environmental regulations affecting a firm and the firm's compliance with those regulations also may influence the environmental value of adopting an EMS. An EMS is likely to have less of an impact in facilities that have already adopted environmental improvements because of regulations than in those that have previously paid little attention to these matters.

Clearly, numerous distinguishing characteristics of firms and their EMSs could lead to different conclusions than those reached by the two studies cited above. More research is needed to refine understanding of EMSs and industry characteristics that would warrant a policy to reward or mandate an EMS. Without such research, using an EMS as a policy tool is premature.

However, this conclusion does not suggest that this sort of process-focused research should be a high government priority. Imagine that government discovered that businesses using a certain cost-accounting method as part of their EMS realized significant gains in environmental performance compared with a control group that did not. A logical policy response would be for government to reward only those EMSs using that cost-accounting method. However, such a response would risk destroying the innovation dynamic of the system. Consultants would have little incentive to develop better EMS tools, because they would have to convince regulators to change their definitions of minimum acceptable EMS process standards before they could market the new method. It is more appropriate for businesses to sponsor process-focused research.

What about the potential value of EMSs for building a more performance-focused, information-driven environmental protection system? As noted earlier, neither the original Responsible Care program nor ISO 14001 ensures a focus on environmental outcomes, even though both encourage pollution prevention. Neither system mandates or encourages public disclosure of environmental performance information or establishes a system of comparable environmental performance metrics. The ISO 14001 standard establishes a third-party certification process for interested firms, but certifiers certify only that a firm's policies and management systems meet the ISO 14001 requirements and follow the firm's own EMS policies and management system. They make no attempt to verify that a company or facility has measured its environmental performance accurately, reported it correctly to the public or regulatory bodies, or improved environmental performance (Morrison and others 2000, 46–47). Thus, mandating or rewarding increased use of EMSs per se will not ensure a more performance-focused, information-driven environmental protection system.

The highly uncertain value of EMSs suggests that policymakers should emphasize the generation of credible and comparable environmental outcome information over efforts to encourage EMS adoption. That is not to suggest that increased use of EMSs might not have environmental value but to underscore that despite the recent upswing in EMS activity, the information "horse" should precede the EMS "cart" as a policy tool. Only two (of 19) state environmental leadership and performance track programs launched in the late 1990s[9] and EPA's National Environmental Performance Track program announced in late June 2000 recognize the need to place primary emphasis on improving environmental performance and generating public performance information. They offer rewards only to those companies that use EMSs that have environmental performance objectives, commit to improved (beyond compliance) environmental performance, and measure and publicly report performance progress. The Performance Track program takes an important step toward strengthening the information infrastructure by basing its framework for reporting about facility environmental performance on an emerging industry standard for reporting the environmental performance of corporations, the Global Reporting Initiative (GRI 1999).

What Does Business Want to Know about EMSs?

Government regulators and policymakers are not the only people interested in learning about the value of using an EMS. Business (as well as government and nonprofit facility and fleet) managers also want to know whether and how using an EMS can help them, particularly whether it will reduce their operational risks, improve their operational efficiency, or help them address regulatory concerns (Conference Board 2000). At least in theory, the use of an EMS could help a business reduce costs, increase revenues, and even improve market position relative to competitors. Similarly, an EMS could help a company reduce exposure to several different kinds of risk: the financial and humanitarian risk of workers or the community being exposed to environmental harm, the time risk of project delays associated with contentious environmental permit reviews, and simply the reputational risk of being in violation of environmental regulations.

A recent study found that many businesses also expect EMS use to improve their relationships with the public. In a survey of 45 companies, the Conference Board (2000, 18) finds that beyond ensuring compliance, businesses rank "corporate image," "access to communities," and "public opinion" as the primary expected benefits of EMS use.[10] Surveys of European and Japanese companies affirm that they, too, hope EMS use will help with external relations (Morrison and others 2000, 51).[11]

Investors and insurers also may be interested in knowing how EMS use affects a firm's financial and risk position. Investors would be interested in learning whether EMS use enhances the financial position of a company. Lenders may want to know whether EMS use enhances a borrower's ability to repay. Those planning to insure other firms are most likely to be interested in the value of EMSs for risk reduction.

Does the information infrastructure exist to meet these various business information needs? To some extent it does. Business, academics, nonprofit organizations, and consultants have prepared numerous case studies that describe environmental management strategies used by businesses and, in some cases, try to assess their impact. For example, some case studies on environmental accounting systems find that EMSs can contribute to cost reduction if appropriately structured (Ditz and others 1995; Epstein 1996). Others show that incorporating the environment into a company's overall management system can boost a firm's financial position by identifying revenue enhancement opportunities (Piasecki 1995; Reinhardt 1999).

Despite the wealth of case studies, business needs to exercise caution in wholeheartedly relying on their findings. Case studies produced by consultants to promote a particular strategy or theory are inevitably plagued by an overly favorable bias. Even case studies produced by businesses themselves may be overly affirmative, because businesses pleased with their EMS experience are more likely to share the story than those that are not. Moreover, case studies often do not capture the particular characteristics of a business or the EMS—either its design or the process used to prepare and implement it—that might affect the value of using an EMS.

Furthermore, whereas case studies can be especially useful for eliciting the characteristics of an EMS that may affect outcomes, they cannot demonstrate that EMS use leads to better environmental outcomes, higher compliance levels, or reduced risk because of the absence of a control group. As with public policymakers, the business sector would benefit from comparable metrics, such as the TRI data King and Lenox (2000) used, that allow comparison of businesses that have adopted an EMS with those that have not. The recent effort by the Environmental Law Institute and the University of North Carolina at Chapel Hill to build the National Database on Environmental Management Systems (NDEMS) holds great promise for strengthening the EMS information infrastructure that businesses need (ELI not dated). NDEMS, a project financially supported by the EPA and enthusiastically promoted by the Multi-State Working Group on Environmental Management Systems (Smoller 1998), is collecting quality-assured, comparable information about EMS characteristics and how EMS adoption affects the environmental and economic impacts of both publicly and privately owned facilities using EMSs. More than 50 facilities voluntarily submit information to the NDEMS.

The NDEMS will provide researchers a valuable and previously unavailable database for analyzing EMS use. The data will greatly enhance businesses' ability to learn about the value of different EMSs. Business would be well-served by contributing information and providing financial support to the NDEMS effort.

The lack of a control group of non-EMS users reporting environmental outcome information limits the ability of the NDEMS to answer the government's fundamental question about whether EMSs improve environmental performance. However, the NDEMS can provide useful descriptive information that helps state and federal regulators understand how their regulations affect individual businesses (ELI 2000). Government took an important and commendable step in sponsoring the NDEMS effort; it should continue to support this piece of the environmental information infrastructure financially, and business should begin to contribute routinely, laying the foundation for the sort of public–private information infrastructure partnership that so dramatically boosted agricultural productivity.

If businesses want to realize their goal of using EMSs to improve their public image and relations, they also need to pay far more attention to making sure EMSs deliver credible and comparable environmental performance information to the public, a subject explored further in the next section.

What Do Citizens Want to Know about EMSs?

Citizens are unlikely to have much interest in EMSs per se, just as they have little interest in other sorts of management systems that businesses use. However, they are likely to care about how the activities of a business might affect them and their families, friends, and neighbors. They want to know whether a business is exposing them to environmental, health, and safety risks. They want to know whether the information a business provides can be trusted. They want to know that the business has been and will be appropriately penalized for violation of environmental laws. They may also want to know how a business contributes to environmental improvements in the community. Thus, as with government, what citizens need is credible information about environmental performance, presented in a manner that allows them to interpret it.

With regard to compliance with environmental laws, the ISO 14001 certification established to verify that companies have a system in place to ensure compliance with environmental laws and improve their environmental management processes is still too weak for the public to trust (Morrison and others 2000). Regarding environmental impact, few companies publicly report environmental outcome information. That which is reported is reported differently by each company (Lober 1997). This lack of standardization makes the information hard for the public to interpret and put in context. Interpretation is facilitated by translating and packaging information in a context familiar to the audience. As noted earlier, standardized metrics that allow comparison with similar companies, prior performance, or other familiar situations provide that context. The use of standardized protocols for reporting measured performance would greatly boost their value to the public. The credibility would be further boosted if private firms retained verification services from a respected "arm's length" third party, such as a nonprofit environmental group.

EMSs could contribute to the information infrastructure needed to provide citizens with the information they need. They will not necessarily, and in most cases do not currently, do so. EMSs that generate a broader set of credible and comparable facility-specific environmental performance information would greatly benefit citizens because they would enable neighbors to understand local impacts. Anything less is unlikely to have the benefit citizens—and companies hoping to improve community relations—seek.

What Are the Essential Elements of a Robust Environmental Management Information Infrastructure?

The preceding discussion begins to suggest how EMSs can contribute to a robust environmental management information infrastructure that can support a performance-focused, information-driven regulatory system.[12] In the following discussion, I briefly explore several key issues associated with meeting these various information needs.

A Broad Set of Performance Measures

To assess the effectiveness of environmental management strategies and present a full and accurate picture of a business's impact on the environment, we need a wide array of facility-specific performance measures. One possible list is suggested here. This list may not be inclusive but begins to give an accurate picture of the impact of a business or other source on the environment:

- *toxic releases,* as an indicator of community exposure levels
- *accidental releases,* as an indicator of community exposure levels and of the safety of internal materials handling procedures
- *release of nontoxic emissions* to the air, water, and ground, as an indicator of threats to ecosystem balance
- *toxics use,* as an indicator of the risk of community and worker exposure
- *nonrenewable (virgin) resource consumption,* as an indicator of the depletion of Earth's (and society's) assets
- *solid waste generation,* as an indicator of land use impacts and future cleanup costs that will be borne by the community
- *noncompliance,* as an indicator of fairness and the fulfillment of social obligations essential to a law-abiding culture

Despite the current limited availability of most of these measures, gathering them is feasible and worthwhile. As discussed earlier, TRI provides one dimension of performance information: toxics released. Other databases exist that might be tested for their value in measuring other dimensions of environmental performance. Water and utility suppliers, for example, already track water and electric use by facility. Several states collect toxics use data that could be studied. Several companies also are beginning to demonstrate the technical, economic, and political feasibility of generating and making public a full range of environmental performance measures. Interface, Inc., did this, for example, in its *Sustainability Report* (Interface, Inc., not dated). If companies that adopted EMSs followed the example set by Interface and publicly reported multidimensional environmental performance measures for the company as a whole (and, going beyond what Interface did, for individual facilities), business and government would gain improved understanding of the effectiveness of different environmental management strategies. It also would give a more accurate picture of the net impact of a business or facility on the environment and would improve citizen understanding. EPA, the U.S. Congress, states (through the Multi-State Working Group on Environmental Management Systems or other associations), government facilities, business associations interested in pursuing self-regulation, and individual companies hoping to advance the viability of more flexible strategies to advance environmental protection should move to adopt a much broader set of standardized environmental measures that gauge performance along multiple dimensions. It not only would support better decisionmaking but also is essential and would accelerate efforts to focus on corporate environmental outcomes rather than corporate adherence to process requirements.

Credibility of Measures

Performance measures and case studies have no value if they cannot be trusted. Consumers trust the performance information they read in *Consumer Reports* because an objective external party produces it and because they have found past analyses to be valid. Consumers are far more wary of product performance information provided solely by manufacturers.

Despite the likelihood that self-reported performance information presents a more favorable picture for the performer than externally produced information, the environmental protection system in this country makes extensive use of self-reporting with post hoc government oversight of reporting accuracy. Unfortunately, there is little sense of how well this approach works, a problem that not only impedes the ability to assess program effectiveness but also affects fairness across regulated parties as well.

One way some government agencies have tried to adjust for insufficient government resources for reviewing the actions of private parties is by deputizing for-profit or not-for-profit private parties to conduct oversight to ensure data quality. The preeminent EMS system, ISO 14001, allows companies to purchase the services of registered external parties to review and certify that a company's EMS meets ISO standards. One advantage of using third-party reviewers (funded by the parties being reviewed) is that they can more easily expand or contract relative to the demand for reviews than can the public sector, which needs to fight recurring battles with budget offices and legislators for every increment of new staffing.

Deputized and self-review systems only work when someone "checks the checkers" and punishes lax reviewers and self-reporters sufficiently to motivate honesty and competence. As noted earlier, questions have been raised about the quality of the reviews by the new ISO registrars. Concern has been expressed that most ISO 14001 registrars have experience with quality management systems, not EMSs, and therefore may not be qualified to detect environmental problems. In early 2000, the National Academy of Public Administration launched a study of the ISO 14001 EMS conformity assessment process (NAPA 2000).

An ongoing process to ensure information credibility is an essential and critical component of a healthy environmental management information infrastructure. Therefore, it should become a core concern that both the states and EPA address. Putting this sort of system in place at EPA will require leadership by the U.S. Congress to assign organizational responsibility and allocate sufficient funding. If Congress fails to step up to this challenge in a timely manner, states should fill the void. And if government fails to step forward, industry should pressure government to do so.

At the same time, industry associations and individual businesses experimenting with voluntary self-regulation systems should establish their own credibility-assuring capacity as part of such systems. For example, businesses could sponsor such respected third parties as the Environmental Defense Fund, the Natural Resources Defense Council, or the World Resources Institute to verify the accuracy of publicly reported environmental performance information.

Finally, research on third-party oversight systems and other credibility-assuring mechanisms, in and beyond the environmental field, would be useful. As noted earlier, one such system undergirds public financial markets. Certified public accountants (CPAs) ensure corporate compliance with generally accepted accounting principles and the reporting requirements established by the U.S. Securities and Exchange Commission. CPA performance, in turn, is monitored by the government and the marketplace. Both private parties (through the court system) and government agencies enforce compliance with those standards.

Research is needed to understand when and why third-party oversight systems work. Such research should examine professional standards, mandatory training, how audit frequency and intensity affect compliance levels, government oversight of checkers, effective incentive systems, and the importance of public reporting. People and industries interested in developing performance-focused, information-driven systems should sponsor, lobby for, and otherwise support research in this area.

Context for Interpreting Information

Standardization and normalization within a company allow the comparison of performance from year to year and across organizational units. The ability to compare across units allows companies to identify their best performing units for benchmarking and for smarter resource allocation decisions. It enables some companies to compare environmental cost-accounting information to decide which plants should manufacture specific products (Ditz and others 1995, 27). Standardization and normalization across companies provide even greater value. They allow companies to benchmark with each other and citizens to interpret corporate environmental performance information.

Standardizing environmental performance measures will greatly boost efforts to assess the effectiveness of different environmental management strategies. Despite the flaws in the TRI, the fact that EPA standardized the method for measuring toxics releases enabled King and Lenox (2000) to conduct their study (after they used standardized commercial databases and extensive calculation to normalize the information). However, the TRI is only one database. Absence of information standardized across companies makes it impossible to assess the impact of an EMS or other intervention strategies on additional dimensions of environmental performance.

A standardized set of measures is essential to establishing a fair performance-focused system that will let strong environmental performers—not those with the best marketing apparatus—enjoy special privileges.

The U.S. Congress, EPA, and the states should move rapidly to experiment with and eventually expand requirements to report standardized environmental performance measures. However, it is not necessary to wait for mandatory governmental standards. Businesses and other facilities can voluntarily agree to report additional environmental information in a standardized format. Many businesses and local governments are voluntarily participating in NDEMS. Others are joining the GRI and could agree to apply the measures GRI is developing for corporations to individual facilities. Business leaders hoping to create a strong environmental protection system that affords greater flexibility than the current system could accelerate the transition to such a system by supporting and participating in cooperative efforts to generate credible, comparable environmental performance information.

Meeting users' needs requires more than just the generation of credible and comparable performance information. Raw data must be analyzed and translated into answers to users' questions and presented in a format and language users can readily understand. Who will do this transformation and translation? Historically, federal support for research on the effectiveness of different environmental management intervention strategies has been sparse and erratic. In the past few years, Congress and a few foundations have begun to earmark a small amount of funding for the evaluation of management experiments, an effort that warrants extension and expansion. State financial support for this sort of analysis has been just as scanty. Businesses and trade associations could further bolster the accumulation of knowledge by sponsoring objective, peer-reviewed research on environmental management strategies. A solid information infrastructure to support a dynamic environmental management learning system needs far greater financial support from federal and state governments, business, and the foundation community than it currently gets.

Dissemination

How do policymakers learn about and gain access to the analyses that have been done? An unread report or analysis sitting on a library shelf is like a tree falling in the forest that nobody hears: it doesn't make any noise.

Numerous vehicles exist to move information from one mind to another, including journals, websites, and professional associations. The problem is that environmental management information is simultaneously too broadly dispersed and too hard to find. There is no commonly accepted opinion-leading or news-breaking publication that the interested audience can rely on to filter relevant information. Studies about the effectiveness of environmental management strategies and the environmental performance of different sources can be found in several dozen general management journals, environmental management and law publications, industrial ecology publications, the general and environmental accounting press, environmental news publications, and other sources. Because many publications carry occasional bits of relevant information, public and private policymakers find learning about research relevant to their environmental management decisions a costly and time-consuming endeavor.

Packaging and disseminating information remains another weak link in the environmental management information infrastructure sorely in need of both public and private attention. The filtering and delivery mechanisms have to be improved along with the systems for translating, transforming, and packaging information for the potential citizen, business, and government audience. If a public or private entrepreneur does not step forward to fill this information gap, the federal government, with support from the states, should.

Using Performance Information to Drive Continued Environmental Gain

The goal in building a performance-focused, information-driven environmental protection system is to create a dynamic system that persistently encourages companies and other sources to pursue environmental gain, just as the market rewards firms that continually pursue financial improvement. For information to function as the key driver in this system, more research is needed to understand the mechanisms likely to encourage sources of environmental problems to pursue environmental improvements. Moreover, research already conducted about targets, information, and incentives needs to be found, translated, packaged, and supplemented to help policymakers better understand how to use performance information to motivate improved environmental outcomes.

Conclusion

Performance-focused, information-driven strategies for environmental protection hold great promise for improving environmental outcomes. However, they depend on a robust information infrastructure. That infrastructure must be able to deliver credible and comparable performance information to public and private decisionmakers in government, business, and communities. It must be able to translate and transform that information so that the potential information audience appreciates its relevance and understands how to interpret it. It must be able to deliver the information to users when and where they need it. And it must be able to help the information audience understand strategies likely to motivate improved environmental performance and those unlikely to produce the desired gain.

The existing environmental information infrastructure is weak and requires significant amplification. Little outcome information exists to supplement the environmental performance picture TRI presents. NDEMS and GRI will be important next steps for generating and accumulating comparable performance information if companies adopt GRI at the facility level. ISO 14001 and other EMS systems hold promise, but if and only if they ensure the use and public disclosure of comparable environmental performance information and strengthen the mechanisms that ensure the credibility of that information.

The federal government, states, localities, business, nonprofit organizations, and foundations should direct greater attention and support to building the information infrastructure essential to the success of performance-focused, information-driven environmental protection strategies. Possible tasks include

- testing, adopting, and publicly reporting a set of standardized environmental performance measures much broader than the few currently in use;
- understanding, establishing, and broadly using credibility-assuring mechanisms;
- analyzing performance information to meet the "business" needs of government, business, and the community and disseminating the analyses to the appropriate audiences;
- generating objective case studies that provide a richer understanding of why certain outcomes occur;
- enhancing skills and boosting the capacity to translate information to users in a manner they can understand;
- improving information delivery and filtering systems so that relevant information will reach potential users when they need it, in a form they can use, and at a cost they can afford; and
- developing better understanding of how to use performance information successfully to motivate environmental improvement.

The challenge is to build, support, or otherwise encourage the creation of the institutional capacity to do all of the above on a recurring basis, so that as learning occurs, the system will evolve.

It is neither reasonable nor desirable to expect the information infrastructure to be built in a coordinated manner. Multiple parties—including government, business, and citizens, acting individually and collectively, sometimes competitively and sometimes in support of each other—all need to step forward and make their contributions. Hopefully, motivated by the shared vision of a performance-focused, information-driven system that can deliver a more dynamic, inclusive, and fair policy tool for government; a more effective and efficient learning system for business; greater accountability to the public; and improved environmental quality for everyone, they will.

Notes

[1]In his 1990 book on systems thinking, MIT professor Peter Senge explains the importance of thinking systematically, instead of focusing on snapshots or isolated parts of a system. He describes a system as "bound by invisible fabrics of interrelated actions, which often take years to fully play out their effects on each other…. Systems thinking is a conceptual framework, a body of knowledge and tools that have been developed over the past fifty years, to make the full patterns clearer, and to help us see how to change them effectively." In his 1980 book on competitive strategy, Harvard Business School professor Michael Porter describes strategy as "a broad formula for how a business is going to compete, what its goals should be, and what policies will be needed to carry out those goals … a combination of the *ends* (goals) for which a firm is striving and the *means* (policies) by which it is seeking to get there."

[2]In an effort to support an international environmental management system (EMS) standard, the United Kingdom replaced its standard with ISO 14001 when this EMS standard was created by the International Organization for Standardization in 1997.

[3]The discussion that follows on the effect of goals and information on motivation and performance is informed by a summary of the literature (Katz 2000). Assertions about the relationship among performance, information, motivation, and outcome changes that are not footnoted are just that—my assertions, based on observation and, I hope, logic and common sense.

[4]California's Safe Drinking Water and Toxic Enforcement Act of 1986 is frequently referred to as Proposition 65.

[5]Despite the strengths of the Toxics Release Inventory (TRI) information system, it also has many weaknesses. King and Baerwald (1999) point out that TRI measures all released toxics in pounds; it does not distinguish the varying levels of toxicity or environmental impact per pound of different chemicals. Nor does it distinguish between transfers to treatment facilities and releases to the environment. Moreover, according to a 1997 study by the Environmental Defense Fund (EDF 1997), only 25% of top-volume chemicals in use have been subject to toxicity testing, a necessary first step before a chemical is listed in the TRI reporting system. A few recent changes in TRI law and voluntary actions by chemical manufacturers may lessen some of these problems, but many will remain.

[6]In October 2000, a working group of state and U.S. Environmental Protection Agency (EPA) environmental managers released a blueprint for the design of a new environmental information infrastructure that holds great promise for correcting the data quality problems that currently plague EPA databases (ECOS 2000).

[7]The Responsible Care program is considered by many to be not an environmental management system (EMS) but a code of conduct. Because governments that consider rewarding EMSs have not yet clearly or commonly defined what an EMS is, that distinction is not relevant to the conclusions of this chapter.

[8]King and Lenox (2000) look at 3,606 facilities associated with 1,500 companies that represent 95% of the production volume of the chemical industry. In 1990, 130 companies participated in Responsible Care; in 1996, 160. The study employs regres-

sion analysis using as the outcome measure (dependent variable) TRI data reported by individual chemical manufacturing facilities. For input measures (independent variables) and to normalize the outcome measures so that comparisons could be made over time and across facilities while controlling for changing production levels, the study uses facility characteristics from the Dun & Bradstreet data, the Duns Database, and participation in the Responsible Care program.

[9]Alaska and Oregon require measurement, public communication, and proven "beyond compliance" performance (Crow 2000).

[10]Benefits that ranked lower than corporate image, access to communities, and public opinion were operational efficiency, new business opportunities, and future competitive position. Perhaps most striking, 86% of the 38 companies identified as peer leaders indicated that use of an environmental management system contributed to corporate image, compared with only 56% of the other companies.

[11]A survey of 500 European companies conducted by SGS Yardley International Certification Services Ltd. found that 81% of firms were adopting environmental management systems to comply with legislation, 80% to improve market share, 78% because of customer pressure, and 64% for public recognition. A survey of 323 International Organization for Standardization–registered Japanese firms found that 82% of firms sought an improved company image.

[12]For a fuller discussion of key components of the information infrastructure, see Metzenbaum 1998.

References

Austin, Robert D. 1996. *Measuring and Managing Performance in Organizations.* New York: Dorset House Publishing.

Brandeis, Louis D. 1914. *Other People's Money, and How the Bankers Use It.* New York: Frederick A. Stokes.

Chandler, Alfred D. Jr. 1977. *The Visible Hand: The Managerial Revolution in American Business.* Cambridge, MA: The Belknap Press.

Conference Board. 2000. *Corporate Environmental Governance: Benchmarks toward World-Class Systems.* New York: The Conference Board.

Crow, Michael. 2000. Beyond Experiments. *Environmental Forum* 17(May–June): 19–29.

Ditz, Daryl, and others 1995. *Green Ledgers: Case Studies in Corporate Environmental Accounting.* Washington, DC: World Resources Institute.

E4E (Enterprise for the Environment). 1998. Bi-Partisan Group Announces Recommendations to Improve the Environment. January 15. Press release announcing the release of *The Environmental Protection System in Transition: Toward a More Desirable Future*, final report of the Enterprise for the Environment. Washington, DC: Center for Strategic and International Studies.

ECOS (Environmental Council of States). 2000. Report to the State/EPA Information Management Workgroup. Blueprint for a National Environmental Information Exchange Network. October 30. Washington, DC: ECOS. http://www.sso.org/ecos/projects/ (accessed January 14, 2000).

EDF (Environmental Defense Fund). 1997. *Toxic Ignorance: The Continuing Absence of Basic Health Testing for Top-Selling Chemicals in the United States.* New York: EDF.

———. 2000. *Air Emissions of Specific Toxic Chemicals 1988–1997, Trend for U.S. and California: Carcinogens and Reproductive Toxins.* New York: EDF. http://www.scorecard.org/wip/Air_Trends.htm (accessed April 21, 2000).

ELI (Environmental Law Institute). 2000. *National Database on Environmental Management Systems: The Effects of Environmental Management Systems on the Environmental and Economic Performance of Facilities.* Second Public Report. June. http://www.eli.org/pdf/PR2_FINAL.pdf (accessed October 8, 2000).

———. Not dated. *National Database on Environmental Management Systems.* http://www.eli.org/isopilots.htm (accessed October 8, 2000).

ENDS Report. 1999. Ford, GM Push ISO 14001 Down Their Supply Chains. London, U.K.: Environmental Data Services. http://www.ends.co.uk/report/Oct99_2_tx.htm (accessed September 25, 2000).

Epstein, Marc J. 1996. *Measuring Corporate Environmental Performance: Best Practices for Costing and Managing Effective Environmental Management Strategies.* New York: McGraw-Hill.

Evenson, Robert B., and others 1979. Economic Benefits from Research: An Example from Agriculture. *Science* 205 (September): 1101–1107.

GRI (Global Reporting Initiative). 1999. *Sustainability Reporting Guidelines: Exposure Draft for Public Comment and Pilot Testing.* March. Boston, MA: Coalition for Environmentally Responsible Economies. Updated at http:\\www.globalreporting.org (accessed October 6, 2000).

Gormley, William T. Jr., and David L. Weimar. 1999. *Organizational Report Cards.* Cambridge, MA: Harvard University Press.

Gouldson, Andrew, and Joseph Murphy. 1998. *Regulatory Realities: The Implementation and Impact of Industrial Environmental Regulation.* London, U.K.: Earthscan.

Hoffman, Andrew J. 1999. Institutional Evolution and Change: Environmentalism and the U.S. Chemical Industry. *The Academy of Management Journal* 42(August): 351–371.

Howard, Jennifer A., Jennifer Nash, and John Ehrenfeld. 2000. Standard or Smokescreen? Implementation of a Non-Regulatory Environmental Code. *California Management Review* 42(2): 63–82.

Interface, Inc. Not dated. *Sustainability Report.* Atlanta, GA: Interface, Inc.

Kaplan, Robert S., and David P. Norton. 2001. *The Strategy Focused Organization.* Cambridge, MA: Harvard Business School Press.

Katz, Nancy. 2000. Incentives and Performance Management. Paper prepared for the Executive Session on Public Sector Performance Management, June 19–21, 2000. Cambridge, MA: Harvard University, Kennedy School of Government.

King, Andrew, and Sara Baerwald. 1999. Using the Court of Public Opinion to Encourage Better Business Decisions. In *Better Environmental Decisions*, edited by Ken Sexton and others. Washington, DC: Island Press.

King, Andrew A., and Michael J. Lenox. 2000. Industry Self-Regulation without Sanctions: The Chemical Industry's Responsible Care Program. *The Academy of Management Journal* 43(4): 698–717.

Lober, Douglas. 1997. Current Trends in Corporate Reporting. *Corporate Environmental Strategies* 4(Winter): 15–24.

McDermott, Richard. 1999. Why Information Technology Inspired but Cannot Deliver Knowledge Management. *California Management Review* 41: 4, 110.

Metzenbaum, Shelley. 1998. Making Measurement Matter: The Challenge and Promise of a Performance-Focused Environmental Protection System. October. CPM 98-2. Washington, DC: Brookings Institution, Center for Public Management.

————. Forthcoming. Making Measurement Useful: A Case Study in Performance-Focused, Information-Driven Environmental Protection. In a forthcoming book, edited by Donald F. Kettl. Washington, DC: Brookings Institution.

Morrison, Jason, and others. 2000. *Managing a Better Environment: Opportunities and Obstacles for ISO 14001 in Public Policy and Commerce.* Oakland, CA: Pacifica Institute.

NAPA (National Academy of Public Administration). 1995. *Setting Priorities, Getting Results.* Washington, DC: NAPA.

————. 1997. *Resolving the Paradox of Environmental Protection.* September. Washington, DC: NAPA.

————. 2000. Evaluating U.S. Registration Practices for ISO 14001: Project Overview. January. http://www.napawash.org/napa/index.html (accessed October 1, 2000).

Pham, Alex. 1999. Putting HMOs to the Test. *Boston Globe,* September 10.

Piasecki, Bruce. 1995. *Corporate Environmental Strategy.* New York: John Wiley & Sons.

Porter, Michael E. 1980. *Competitive Strategy.* New York: Free Press.

Reinhardt, Forest. 1999. Bringing the Environment Down to Earth. *Harvard Business Review* 77: 149–157.

Repetto, Robert, and Duncan Austin. 2000. *Pure Profit: The Financial Implications of Environmental Performance.* Washington, DC: World Resources Institute.

Ruckelshaus, William D. 1998. Preface. In *The Environmental Protection System in Transition: Toward a More Desirable Future,* final report of the Enterprise for the Environment. Washington, DC: Center for Strategic and International Studies.

Senge, Peter. 1990. *The Fifth Discipline.* New York: Doubleday/Currency, 7.

Social Investment Forum, Inc. 1997. *Trends: Report on Responsible Investing Trends in the United States.* November. http://www.socialinvest.org/areas/research/trends/1997-Trends.htm. Accessed October 8, 2000.

Sparrow, Malcolm. 1994. *Imposing Duties: Government's Changing Approach to Compliance.* Westport, CT: Praeger.

Smoller, Jeff. 1998. Memorandum from Jeff Smoller, Secretary of the Multi-State Working Group (MSWG), to Researchers Interested in the Multi-State Working Group on Environmental Management Systems. Information packet. August 8. Madison, WI: State of Wisconsin Department of Natural Resources.

Swift, Byron. 1997. The Acid Rain Test. *Environmental Forum* 14(May/June): 20.

U.S. EPA (Environmental Protection Agency). 1999. *Project XL: From Pilot to Practice.* September. EPA 100-R-99-007. Washington, DC: U.S. EPA, Office of Reinvention.

———. 2000. *National Environmental Achievement Track Program Description.* June 26, 2000. http://www.epa.gov/performancetrack. (accessed October 8, 2000).

Wetherill, Virginia. 1997. Memoranda from Secretary Virginia Wetherill to individual division and district directors. In Briefing Materials, Presentation to Governor Lawton Chiles, Florida Department of Environmental Protection, Secretary's Quarterly Performance Report. December 2. Tallahassee, FL.

8

Policies to Promote Systematic Environmental Management

Cary Coglianese

According to a recent National Research Council (NRC 1997, 43) committee report, environmental management systems (EMSs) adopted under international standards offer "the potential to change the culture of companies worldwide on environmental protection and commitment; harmonize environmental management frameworks, labels, and methods worldwide; move companies beyond compliance; and promote voluntary improvements." These systems can provide firms with processes for identifying environmental impacts, setting performance goals, and generating action needed to improve environmental performance even beyond what is currently required under existing regulations. By acting systematically, firms can identify more precisely their current resource uses and seek alternative processes that use fewer or more benign resources, sometimes with resulting cost savings to the firms.

In this chapter, I assume that EMSs lead to better results and consider what, if anything, government might do to promote the use of such management systems. How, in other words, might policy be modified to promote the widespread and responsible adoption of EMSs? As noted in Chapter 1, numerous current efforts to promote the use of EMSs are under way at the state and federal level in the United States (Crow 2000).

My purpose in this chapter is to analyze various public policies to encourage firms to adopt EMSs and make significant improvements in their environmental performance. These options include those that would reduce firms' costs of implementing EMSs, increase firms' benefits from their

181

implementation, and outright require EMS adoption. I argue that each policy option should be assessed according to a range of criteria, including the
likely impact it will have on firms' incentives to achieve environmental
improvements, its administrative feasibility, and its legal and political
acceptability. Public policy also should take into account how the impact of
each option may vary with respect to different types of firms or different sectors of industry.

Even though we do not yet have extended experience with or empirical
research on all of the policy options presented here to support strong conclusions about how well they meet the relevant policy criteria, enough is known
to offer informed hypotheses about the likely degree to which each option
satisfies these criteria. Numerous policy options are available to promote
EMSs, but this analysis suggests that none of them meet all the relevant criteria very well. Indeed, as is often the case in public policy, only a few feasible and politically acceptable options seem to offer at best a moderate
prospect of achieving the policy objective. In this case, short of requiring
firms to adopt EMSs, agencies can take some measured steps to encourage
firms to implement management systems, but we should not expect these
incentives to be the major drivers of EMS adoption by many firms.

Lowering the Costs of EMSs

The first set of policy options would lower the costs of EMS implementation.
In establishing an EMS, a firm confronts the information costs associated
with learning about EMSs and how they can be effectively implemented. It
faces the costs associated with modifying facilities, processes, staff, and
equipment. It also may need to overcome the costs that may arise should
regulators rely on the documents the firm generates through its EMS to
sanction the firm for legal violations. In this section, I address several policy
options designed to mitigate the costs of EMSs.

General Education

Information represents an initial cost that firms face with respect to EMSs.[1]
A firm's managers need to know what an EMS is and how implementing one
might benefit the firm as well as the environment. Government agencies can
take steps to provide general information about these systems and demonstrate overall support for their adoption. Examples of such steps include promoting academic research on the benefits of EMSs and the general dissemination of information in trade and environmental management publications,
conferences, the Internet, and other modes of communication.

Educational efforts of this nature are feasible for government agencies
because they require limited commitments of staff and no changes in exist-

ing regulations. Over the long term, such efforts can perhaps foster a cultural change within the regulatory community that will be conducive to the adoption of EMSs. Firms may be more likely to adopt EMSs if their competitors, environmental groups, and other organizations support their use. Efforts that can demonstrate the benefits of EMSs to a wide range of actors would seem only to help facilitate their use.

In the shorter term, however, a general educational approach may have only a limited impact on firms' behavior. For large firms that already have extensive environmental departments, the value added by this approach will be minimal at best, because many of these firms are already investigating the potential of an EMS. For many other firms, it simply may not be enough just to know that a systematic effort to improve environmental performance might be good for their business as well as the environment. They probably also will need to know that the benefits to their firms will outweigh the costs associated with developing EMSs, and they may need other incentives to convince them that it will be so.

Technical Assistance

Many firms will need specific information about how to develop EMSs for their own operations. Environmental agencies can offset these costs by providing concrete advice—technical assistance—to firms seeking to establish EMSs. Compared with a general education approach, technical assistance provides a greater incentive because it offsets more of a firm's information costs. In doing so, of course, it shifts these costs to government agencies.

Government agencies may be able to provide technical assistance at a lower cost than the firms, especially on compliance issues, but such assistance still comes at a cost. Meaningful technical assistance cannot be provided to all firms that might benefit from it, so agencies must be strategic in how they deploy this approach. Accordingly, the U.S. Environmental Protection Agency (EPA) recently targeted EMS technical assistance and training in particular sectors such as in local government, metal finishing, and biosolids. Assistance to specific firms can be leveraged by using the experience with these firms to develop sector-specific EMS "templates" that other firms can use to adopt EMSs even without directly participating in the technical assistance program.

Technical assistance may be limited by firms' reluctance to invite government personnel into their facilities. To be meaningful, technical assistance may require that government officials visit plants and acquire detailed information about firms' environmental impacts. Yet some firms may fear that government employees rendering technical assistance will discover legal violations that would otherwise go undiscovered.[2] It is unclear how prevalent or strong this "fear factor" is, but it may lead agencies to want to offer some limited enforcement forbearance in conjunction with technical assistance. It also may help if agencies provide technical assistance in partnership with sector-based trade associations, a step that could both alleviate some fear on the part of individual firms as well as leverage the technical resources of government agencies.

Subsidies or Tax Credits

Subsidies or tax credits could in principle be offered to induce firms to establish EMSs, directly compensating them for the cost of implementing EMSs.[3] If subsidies or tax credits are sufficiently high, they can provide a very powerful incentive for firms. However, subsidies also can be too low. For example, even though the Minnesota Pollution Control Agency offered to subsidize one-half of the costs of compliance audits for printers in the state, after four years, only 15 companies (fewer than 1% of the printers in the state) had requested the subsidy (NAPA 1997, 133–134). Although U.S. EPA and some state agencies have adequate resources to offer limited grants to firms in selected sectors to offset the costs of certain projects, any substantial subsidization for a large number of firms would exceed the resources of government agencies.

Tax credits would have the advantage of not requiring additional appropriations. They also are broadly applicable, potentially encouraging firms throughout all sectors to adopt EMSs. Nevertheless, proposals to provide substantial tax credits probably would not find much political support. Even if legislative support could be secured, the relationship between revenue agencies and environmental regulators would need to be sorted out. Neither tax credits nor subsidies therefore appear to be viable options for encouraging the widespread adoption of EMSs.

Audit Protection

In the course of implementing EMSs, firms conduct assessments of their current environmental practices and subject themselves to internal and third-party audits. They have an incentive to adopt such systems to minimize the risks of noncompliance and to achieve any cost savings that may arise from a more efficient management of materials and energy. However, firms also face a disincentive for conducting internal audits of environmental impacts. If the documents produced during these audits show that a firm has failed to comply with prevailing regulatory requirements, such documents constitute admission of violation and may be used by the government or environmental groups to prosecute enforcement actions or citizen suits against the violating firm. The risk that audit documents might later be used against a firm could be considered an additional "cost" of implementing an EMS and may discourage some firms from launching such systems.

In an effort to counteract disincentives associated with the potential release of audit documents, several states have adopted self-audit privilege laws that provide varying degrees of protection to internally created environmental management documents.[4] For example, in Oregon, any environmental audit report is treated as privileged and generally inadmissible in any legal action.[5] However, the privilege does not apply if the firm failed to act promptly to initiate reasonable efforts to rectify violations documented in its audits. EPA has issued its own audit policy that also aims to remove barriers that could keep firms from conducting compliance assessments (U.S. EPA 1995a, 1995b, 1999). The EPA audit policy does not create a privilege for audit documents but does articulate the agency's general (though nonbinding) position that it will refrain from making routine requests for internal audit reports.

It is not clear precisely what impact self-audit policies have in terms of encouraging firms to develop EMSs. Intuitively, such policies would seem to remove a potential "cost" associated with conducting self-audits by offering assurance that regulators will not use audit documents against firms. However, firms may already have sufficient incentive to conduct audits notwithstanding any potential risk that regulators will use audit documents against firms. In a recent evaluation of its audit policy, EPA acknowledged several studies that reportedly show that the overwhelming incentive for firms to implement voluntary audits is to correct noncompliance problems before government inspectors discover them (U.S. EPA 1999).[6]

Moreover, it is less clear what effect audit policies will have on the development of broader EMSs. In its evaluation, EPA cited some evidence suggesting that its audit policy encouraged firms to implement EMSs. Of 50 firms surveyed that had reported violations under EPA's audit policy, about one-half reported that they had in place either an EMS or a due-diligence compliance management system. Of these, half reported that the EPA's audit policy had "encouraged specific improvements" in these systems (U.S. EPA 1999).

A central question remains with respect to the number of firms that would develop such systems in the absence of policies that protect the confidentiality of internal audit reports. Such policies do appear to mitigate a disincentive for creating management systems that include compliance audits, but they may ultimately provide only a moderate impetus for firms to develop EMSs in the first place.

Increasing the Benefits of EMSs

Just as lowering the costs of EMSs could help promote their use, so too could efforts to increase the benefits that accrue to firms that adopt these systems. Government agencies could potentially offer firms preferential treatment in the form of public recognition, enforcement forbearance, and regulatory and permitting flexibility. These incentives are all possible ways of increasing the private benefits to firms for implementing EMSs.

Public Recognition

Initiatives designed to promote voluntary environmental efforts have frequently offered public recognition to firms. EPA's 33/50 and Energy Star programs, for example, have offered firms various types of recognition. Public recognition can take the form of certificates of participation, product labeling, and even government-sponsored publicity. Recognition gives firms a distinction that they can use to differentiate their products and demonstrate to employees and local communities that they practice exemplary environmental stewardship.

By itself, public recognition will not provide a major incentive for most firms to adopt EMSs, except perhaps in a few industries where product differentiation on environmental grounds adds significantly to a firm's competitive posture. As a result, few programs have offered only public recognition. EPA's 33/50 program was an exception, but then participation in this program also was quite exceptional in that it demanded little from firms other than submitting a pledge to try to reduce emissions of specified chemicals. When EPA tried to establish a second phase of the 33/50 program, industry reportedly balked because the proposed second phase would have imposed greater demands on it (Davies and Mazurek 1996). To encourage firms to make substantial voluntary commitments, something more than public recognition will be needed.

Still, public recognition has one major advantage: it is extremely easy for government agencies to offer. It costs the agency little and demands no changes in existing regulations. For this reason, agencies usually offer public recognition in conjunction with other benefits, and we are likely to see public recognition incorporated into initiatives designed to promote the use of EMSs. The main drawback to including public recognition in a package of incentives is that its effectiveness may diminish with increased use. To the extent that public recognition works, it works because it offers firms a mark of distinction. If agencies offer public recognition on a routine basis, recognition may become something that is expected, and its impact could be reduced.[7]

Enforcement Forbearance

Another step agencies could take to encourage the use of EMSs would be to modify their use of enforcement discretion.[8] Rather than immediately imposing fines, regulators may work with firms to correct the violations and improve their EMSs. They may decline to take action against violations that were disclosed through a firm's EMS and corrected in a timely manner. The StarTrack program in EPA Region 1, for example, provides a 60-day compliance correction period for violations discovered by firms. Several other EPA policies rely on enforcement forbearance for certain violations disclosed by firms, including the agency's audit policy, its small business compliance incentives policy, and its Toxic Substances Control Act (TSCA) enforcement response policy (Hartman and Raclin 1994). Similarly, state agencies offer limited enforcement forbearance to encourage audits and EMSs.

The benefits of enforcement forbearance are limited in at least two ways. First, most such policies limit forbearance to the less serious civil violations. These policies typically neither immunize firms from criminal penalties nor prevent agencies from taking enforcement action in cases that involve significant and imminent endangerment of public health. Second, forbearance policies adopted by government agencies provide no protection against citizen suits. Environmental organizations still could bring actions against firms, even if the government does not. For these reasons, the overall impact of enforcement forbearance on firms' decisions to adopt EMSs will probably be moderate at best.

Regulatory Flexibility

A potentially stronger incentive would be for government agencies to authorize changes to the regulations that govern a firm's operations, including changes to the permitting process. Firms that have exemplary environmental practices would be rewarded with flexibility in achieving their environmental goals. Firms could be allowed to make internal trades across media, make improvements in unregulated environmental impacts in exchange for flexibility over regulated matters, or make changes in current permitting or reporting practices. U.S. EPA and some states currently are moving forward with programs that create a "performance track" (sometimes called a "green tier" or "performance ladder") for those firms that consistently comply with and exceed environmental standards.[9] Firms in such a performance track can be granted waivers from permitting and other requirements. As long as waivers do not conflict with statutes (or, in the case of state agencies, do not conflict with federal regulations), firms will not risk citizen suits because the regulations governing the firm's conduct have been changed.

The key challenge with "performance track" programs will be to decide how much extra environmental performance a firm needs to deliver to gain different types of regulatory flexibility. The environmental community and the public probably will continue to demand a showing of superior performance before an agency decides to waive regulations. Initially, this choice of who deserves a waiver may be made on a case-by-case basis, but eventually, agencies may find themselves contemplating a more rigorous codification of the criteria for entry into a performance track. Administrative and other transaction costs will pose the greatest challenge for any performance track program: overcoming complex and sometimes burdensome administrative procedures to decide whether firms are eligible for regulatory flexibility. As simple as it is to declare that responsible companies who consistently exceed environmental requirements ought to receive special treatment, putting this notion into policy will always turn out to be more complicated.

For example, recently the Oregon Environmental Quality Commission adopted regulations to create a Green Permit program.[10] The program provides for a four-tiered system of permits, three of which recognize firms that adopt EMSs and give them increasingly greater flexibility based on their level of environmental performance. The simple principle of rewarding better-performing Oregon firms takes shape in 15 pages of regulations, the first page of which begins by defining 23 terms. Applicants to the program need to demonstrate that they meet specific criteria outlined by the regulations. If the agency decides to accept an application, it must first proceed through notice-and-comment procedures and hold a public hearing, if requested. In addition, the state may need to seek approval from EPA, because the state agency can grant flexibility only in state-imposed requirements (and then only those that would not trigger a revision in a state implementation plan). Along the way, a "meter" is running, and the applicant is obligated to reimburse the state for its costs of processing the application and conducting the approval proceedings (presumably, even if the application is ultimately denied). The state agency, in turn, must provide the applicant with monthly statements of the costs it incurs. And these are the paperwork requirements associated with only the request for regulatory relief. A company's EMS itself will generate paperwork, as will audit inspections and the certification process.

Many years ago, Bardach (1982, 316) predicted that "the more that on-site visits are replaced by a regime of 'self-regulation' ... the more paperwork is likely to increase and multiply." Although technological advances (such as continuous emissions monitoring and the Internet) may be reducing the costs of monitoring and information, the costs of gaining access to a performance track probably will deter many firms from participating. Experience with some of EPA's regulatory reinvention initiatives (such as Project XL) suggests that crafting individualized plans for specific facilities or firms is resource intensive (Blackman and Mazurek 1999). To date, programs involving regulatory flexibility have been limited to a few firms. Only four firms were involved in the pilot testing of Oregon's Green Permit program; about a dozen have participated in EPA Region 1's StarTrack program; and participation in Project XL has been much lower than originally intended. Each of these programs requires that participating firms first explain why they should be part of the program.

Transaction costs can not only discourage firms from participating in site-specific programs for regulatory flexibility but also place demands on regulatory agencies. Although agencies can accommodate modest levels of participation using existing resources, in the event that significantly more firms seek to participate in a performance track, agencies will confront substantial administrative demands in deciding which firms to admit. However, based on recent experience, it does not appear that many firms will participate in performance tracks that purport to grant significant regulatory flexibility.

Although it may seem only fair that those seeking regulatory waivers bear the burden of demonstration,[11] any nontrivial requirements for entry into the performance track probably will deter firms from participating (Davies and Mazurek 1996). This is not to say that regulatory waivers and performance track programs are not worthwhile; they may well be perfectly justified even if they are used only infrequently to correct for the occasional gross inefficiency that arises from the uniform application of rules. Yet government agencies should probably pause before adopting performance tracks for the purpose of expanding EMS use. It may well turn out that performance tracks attract mainly those firms that are already environmental leaders. Moreover, as long as significant administrative hurdles remain to be overcome to enter a performance track, such proposals probably will not on their own lead many new firms to engage in rigorous environmental management.

Mandating EMSs

Policy options to reduce the costs or increase the benefits of EMSs treat firms' decisions to use these systems as voluntary. A different approach altogether would be to mandate that firms implement EMSs. Mandates could be imposed either by government in the form of regulations or by large manufacturers in the private sector who make the existence of an EMS a contractual condition in supplier arrangements.

Public Mandates

Although EMSs are currently conceived as alternatives to conventional regulation, they could in principle be incorporated into public mandates. In other fields of regulation, such as securities, banking, and food safety, government agencies require operational procedures comparable to management and auditing systems. Yet in the environmental arena, the notion of requiring EMSs has gained relatively little attention. Perhaps the only context in which regulators have required the establishment of environmental management and compliance systems has been in the context of settlements of enforcement actions.

Even though a mandatory approach (with appropriate sanctions for non-compliance) has not garnered much support, it has the potential for leading many firms to use EMSs. However, public regulation also may lead to the possibility discussed in Chapter 1, namely, that many firms will adopt EMSs begrudgingly, implementing them in only a token or ritualistic manner. If the overall policy goal is to create a cultural change so that firms diligently and continually look for ways to improve their environmental bottom line, public regulation may not prove effective. Mandating EMSs might dramatically increase the number of firms using these systems without necessarily increasing the number of firms using them effectively.[12]

The specter of regulations and possible sanctions for noncompliance also may cause resistance from firms, especially from those whose managers do not see the need for such systems. Firms may perceive EMS requirements as simply another unreasonable regulatory burden imposed by government and may react by complying with the rules only minimally (Bardach and Kagan 1982). In this way, requiring EMSs could actually undermine important motivational factors that lead firms to make environmental improvements. On the other hand, if EMSs tend to take on a life of their own—such that even firms that begrudgingly undertake them soon see how beneficial they can be—then regulation might be a sensible way to get firms to overcome their initial resistance. However, before government requires EMS use, more research will be needed to explain their role in leading firms to make environmental improvements vis-à-vis other factors that affect environmental performance.

Private Mandates

Although not really an option for public policy, private mandates hold significant potential for increasing EMS use.[13] General Motors and Ford Motor Company recently announced separate decisions to require all their parts suppliers to implement International Organization for Standardization (ISO)-certified EMSs by 2003. Other manufacturing firms are imposing or contemplating imposing similar requirements, creating the prospect that EMS use will spread throughout entire supplier chains in various sectors. Facing the risk of losing their purchase agreements, suppliers will likely respond by adopting certifiable systems.

Private mandates raise some of the same issues that arise with public mandates: they do not, and probably cannot, mandate the diligence and commitment that it may take for firms to make significant environmental improvements. However, it is possible that private mandates will not generate the same kind of active resistance that public mandates could generate. Moreover, private mandates have the advantage—at least from the standpoint of the government—of being easy to implement.[14]

Unlike public mandates, private mandates cannot apply to all firms but will be limited to the firms in any given supply chain. They also have limited applicability when it comes to service firms such as dry cleaners and printers. However, the limited scope of private mandates can be a significant advantage for the purpose of conducting research on EMSs. The environmental outcomes of supply firms that implement EMSs under a private mandate can be compared with similar firms that lack EMSs because they operate in a different supply chain where no mandate exists. In this way, mandates that affect suppliers in one supply chain may create a natural experiment that would allow researchers to answer with greater confidence the question of whether EMSs yield better results.

Assessing the Options to Promote EMSs

In deciding whether to adopt policies that lower the costs or increase the benefits of EMSs, or whether to require such systems, policymakers will need to compare how each option fares in terms of creating incentives for firms to improve their environmental performance. In addition, each option should be compared on the basis of other criteria, including the scope of the option (that is, how many firms it would encourage), whether the option would require new legislation, the political support (or opposition) that the option would elicit, the burdens the option would place on firms, and the administrative feasibility for the government.

In this chapter, I have discussed predictions about how nine different policy options would fare when evaluated against these criteria. Table 8-1 is a summary of testable hypotheses about the relative impacts each policy option would have on the various criteria I have just enumerated. These predictions are qualitative and relative, because most of the policies are still new, and a few are virtually untried. As government agencies experiment further with these various options, they should plan to conduct systematic empirical research to test for the effects of each along several criteria such as those I have outlined here.

If my hypotheses about the various policy options are correct, none of the options will fully satisfy all the relevant policy criteria. However, four options stand out as having the potential to create at least a modicum of success: technical assistance, audit protection, enforcement forbearance, and private mandates. Each of these options could offer at least "moderate" incentives to firms while being legally and politically acceptable and at least moderately feasible. The other options either do not offer even moderate incentives to firms (education, public recognition) or face significant challenges in implementation (subsidies, regulatory flexibility, public mandates). Although the conclusions offered here to facilitate future research are tentative at best,

Table 8-1. Hypothesized Impacts of Policies to Encourage EMSs

Policy	Degree of incentive	Scope of incentive	Legal acceptability	Political acceptability	Feasibility for firms	Feasibility for agencies
General education	○	◐	●	●	●	●
Technical assistance	◐	○	●	●	◐	◐
Subsidies/tax credits	●	●	○	○	●	○
Audit protection	◐	●	◐	◐	●	●
Public recognition	○	◐	●	◐	●	●
Enforcement forbearance	◐	◐	●	◐	●	◐
Regulatory flexibility	●	○	◐	◐	○	○
Public mandates	◐	●	○	○	◐	◐
Private mandates	◐	◐	●	◐	◐	●

Notes: ● = high, ◐ = moderate, and ○ = low.

they nevertheless illustrate how structured analysis can permit decisionmakers to make comparisons among different policy options.

Two caveats are in order in evaluating the competing options. First, I have tended to generalize across all sectors of the economy and across all types of firms in predicting the effects of each policy option. The effects of at least some of the policy options probably will vary for different kinds of firms. The requirements for entry into performance tracks, for example, will be less of an obstacle for large firms than for small firms. Similarly, some kinds of firms probably will respond more to public recognition or to offers of technical assistance than others will. Policymakers must be attentive to these differences.

Second, I have not attempted to explore the impacts of combined policy options. Agencies typically offer these options in bundles. Audit protection policies, for example, usually combine what I have treated separately as audit protection and enforcement forbearance. Performance track programs have offered (or proposed to offer) combinations of technical assistance, enforcement forbearance, regulatory flexibility, and public recognition. From the standpoint of government agencies, it is certainly understandable that they would try to offer all the available incentives to try to encourage firms to make environmental improvements. However, from the standpoint of the analyst, combinations make it difficult to discern whether individual policy options make a difference.

Consider a hypothetical situation in which a government agency offered technical assistance only in conjunction with a performance track. If this combined policy of technical assistance and regulatory flexibility failed to encourage many firms to adopt effective EMSs, we could not infer that technical assistance on its own would not work. It is conceivable that the administrative burdens associated with applying for the performance track combined with firms' reluctance to invite government employees into their plants created sufficient costs to deter firms from participating. However, it would not mean that on its own the strategic use of technical assistance, perhaps in cooperation with a trade association, would not have yielded better (perhaps substantially better) results. In short, policymakers need to be mindful when they combine policy options.

Conclusion

I began this chapter with the reminder that we need to study the impact of EMSs not only to decide whether these systems are worth promoting but also to understand why they work. EMSs can "draw in" employees within a firm, signaling to them that they ought to give a higher priority to reducing the firm's environmental impacts. The system may be designed to motivate employees to look for innovative ways of improving environmental performance or lowering costs. These systems also may provide an institutional mechanism by which a firm's top management can entrench its commitment to environmental goals, independent of how strong or weak these goals may be. The system, in this sense, would "lock in" the firm to achieving environmental gains. This lock-in might be further enhanced through a rigorous system of verification or auditing by qualified third parties. By providing an institutional structure for the achievement of environmental objectives, management systems may minimize slippage from environmental goals over time.

However, the effectiveness of an EMS may depend on something other than the system itself, such as the commitment by a firm's management to making environmental improvements. After all, something other than the management system itself leads a firm to adopt EMSs, such as the managers' desire to avoid liability for noncompliance with regulations, to reduce the costs of resources or energy, or to differentiate the firm as an environmental leader. Once we recognize the factors other than EMSs that lead firms to adopt EMSs in the first place, we can see that these same kinds of factors can be working more directly to affect firms' overall environmental performance, even after an EMS is in place. A management system may simply be a vehicle that firms with a high commitment to environmental performance use to make improvements, and factors such as top management's priority on environmental performance may ultimately be more vital in determining a firm's performance than the mere existence of an EMS. Therefore, even with an EMS, firms can be expected to achieve better environmental performance if their managers are determined to implement their EMSs rigorously and to make investments needed to change production processes or otherwise improve environmental performance. The motivation of top management—not the presence of an EMS—could well be what best explains a firm's environmental performance.

The answer to what explains firms' environmental performance, particularly what role EMSs play in affecting that performance, is far from self-evident. But such answers are needed to craft sensible public policy. If EMSs lead to better results, then it will be appropriate for government to seek ways of encouraging their widespread use. On the other hand, the success attributed to EMSs may, as suggested, depend on attributes of firms that are independent of the management system per se, such as management commitment to environmental improvement. In that case, policy would need to foster management commitment instead of only the formal adoption of EMSs. Policies that merely increase the use of EMSs may do little to encourage the sustained commitment needed for firms to make ongoing environmental improvements.

Therefore, public policy may need to encourage the kind of sustained commitment and diligence that makes EMSs effective instead of merely encouraging firms to create minimally certifiable systems. For those firms that are already environmentally sensitive, perhaps little in the way of public policy will be needed to encourage these firms to exploit the potential of EMSs, because they are already beginning to do so on their own. The bigger challenge will be to encourage the earnest implementation of EMSs by firms that have yet to recognize the potential benefits of EMSs for their businesses and the environment.

This challenge may prove difficult to accomplish. I have evaluated various options to encourage the use of EMSs based on how strongly they would encourage firms to improve their environmental performance, the number of firms they would encourage, their legal and political acceptability, and their feasibility for firms and government. Measured against these criteria, none of the options is perfect. However, at least four options hold the prospect of feasibly creating at least moderate incentives for EMS implementation: technical assistance, audit protection, enforcement forbearance, and private mandates. The other options either do not appear to offer substantial incentives or would likely present significant difficulties in terms of feasibility.

Acknowledgements

The author gratefully acknowledges helpful comments on earlier versions of this chapter provided by Marian Chertow, Jim Horne, David Lazer, Henry Lee, Richard Minard, Jennifer Nash, and Vicki Norberg-Bohm.

Notes

[1]Information costs are particularly noteworthy in this setting because the benefits that a firm receives from its environmental management system (EMS) are ordinarily known only to those within the firm. If the results of an innovation such as an EMS are hard for those outside an innovative firm to see, others will be less likely to adopt the innovation (Rogers 1995, 16).

[2]This fear of discovery may be justified. In Massachusetts, initial compliance audits of 18 firms participating in a demonstration project involving technical assistance found that only 33% of the firms had fully complied with all regulations. Two firms were dropped from the project because serious legal violations were found (Massachusetts Department of Environmental Protection 1997).

[3]I recognize that subsidies and tax credits could just as easily be considered a method of providing firms with new benefits rather than lowering the costs of adopting EMSs. I include them in the section on lowering costs only because I find it implausible that the option would gain any support if it is framed in terms of "rewarding" firms

with tax credits or subsidies. Such an option only stands a chance of being implemented, it seems to me, if it is framed as a means of offsetting or reducing some of the initial start-up costs associated with EMSs.

[4]As of mid-1994, the Coalition for Improved Environmental Auditing reported that about 15 states either had adopted or were considering adopting legislation to protect internal audit documents (unpublished observations by John L. Wittenborn and Stephanie Siegel, 1994). For more discussion of self-audit privilege policies, see Harris 1996, Gish 1995, and Orts and Murray 1997.

[5]*Oregon Revised Statutes*, Title 36, § 468.963, 1997.

[6]The U.S. Environmental Protection Agency (EPA) cited a 1995 Price Waterhouse survey and a 1998 report by the National Conference of State Legislatures indicating that compliance was the primary incentive for implementing an internal audit system.

[7]It is conceivable that, under certain circumstances, public recognition for environmental performance could become a norm of doing business. Much like the Underwriters Laboratories label arguably has become the norm for consumer product safety, "normal" public recognition could generate pressures for conformity with the requirements needed to secure the recognition.

[8]Deciding which cases to prosecute has long been a matter committed to agency discretion, so agencies generally do not need legislative authorization or to follow any special procedures to shift enforcement priorities (see *Heckler v. Chaney*, 470 U.S. 458, 1983). However, if federal agencies reallocate their enforcement authority in such a way as to require firms to go beyond compliance with existing law to gain some enforcement forbearance, they may be required to develop such policies using notice-and-comment rulemaking procedures (*Chamber of Commerce v. U.S. Department of Labor*, 174 F.3d 206, D.C. Cir. 1999).

[9]Project XL of EPA is perhaps the most prominent program that provides regulatory flexibility in exchange for a demonstration of superior environmental performance. EPA also piloted a performance ladder approach in Region 1 with the StarTrack program and recently established its National Environmental Performance Track program.

[10] Requirements for Green Permits, Oregon Administrative Rules 340-014-0100 to -0165 (August 13, 1999).

[11] Donald Elliott (1997:184) has argued that "those who benefit from more flexible compliance [should] have the burden of clearly measuring and documenting that the alternative system delivers better results than the traditional mechanisms it has replaced, using government-approved validation mechanisms."

[12]A sharp increase in the number of firms seeking to implement EMSs could potentially overwhelm the current capacity for qualified third-party verification, thereby reducing the effectiveness of one of the critical methods for ensuring the credibility and effectiveness of EMSs.

[13]Government policy could, of course, seek to encourage leading manufacturers to adopt private mandates.

[14]As with public mandates, a sudden increase in firms seeking to certify their EMSs could place a significant strain on the capacity of the current system of certification or auditing. See note 13.

References

Bardach, Eugene. 1982. Self-Regulation and Regulatory Paperwork. In *Social Regulation: Strategies for Reform*, edited by Eugene Bardach and Robert A. Kagan. San Francisco, CA: Institute for Contemporary Studies.

Bardach, Eugene, and Robert A. Kagan. 1982. *Going by the Book: The Problem of Regulatory Unreasonableness*. Philadelphia, PA: Temple University Press.

Blackman, Allen, and Janice Mazurek. 1999. The Cost of Developing Site-Specific Environmental Regulations: Evidence from EPA's Project XL. Resources for the Future (RFF) Discussion paper 99-35-REV. April. Washington, DC: RFF.

Crow, Michael. 2000. Beyond Experiments. *Environmental Forum* May/June: 19–29.

Davies, Terry, and Jan Mazurek. 1996. Industry Incentives for Environmental Improvement: Evaluation of U.S. Federal Initiatives. Report to the Global Environmental Management Initiative (GEMI). Washington, DC: GEMI.

Elliott, E. Donald. 1997. Toward Ecological Law and Policy. In *Thinking Ecologically: The Next Generation of Environmental Policy*, edited by Marian R. Chertow and Daniel C. Esty. New Haven, CT: Yale University Press.

Gish, Peter A. 1995. The Self-Critical Analysis Privilege and Environmental Audit Reports. *Environmental Law* 25: 73.

Harris, Michael Ray. 1996. Promoting Corporate Self Compliance: An Examination of the Debate over Legal Protection for Environmental Audits. *Ecology Law Quarterly* 23: 663.

Hartman, Barry, and Linda Raclin. 1994. A Primer on Environmental Auditing. White paper. July. Washington, DC: National Legal Center for the Public Interest.

Massachusetts Department of Environmental Protection. 1997. *Evaluation of the Environmental Results Program Demonstration Project*. November 13. Boston, MA: Massachusetts Department of Environmental Protection.

NAPA (National Academy of Public Administration). 1997. *Resolving the Paradox of Environmental Protection: An Agenda for Congress, EPA, and the States*. Washington, DC: NAPA.

NRC (National Research Council). 1997. *Fostering Industry-Initiated Environmental Protection Efforts*. Washington, DC: NRC, Committee on Industrial Competitiveness and Environmental Protection.

Orts, Eric W., and Paula C. Murray. 1997. Environmental Disclosure and Evidentiary Privilege. *University of Illinois Law Review* 1997: 1.

Rogers, Everett. 1995. *Diffusion of Innovations* (Fourth Edition). New York: Free Press.

U.S. EPA (Environmental Protection Agency). 1995a. Voluntary Environmental Self-Policing and Self-Disclosure Interim Policy Statement. April 3. *Federal Register* 60: 16875.

———. 1995b. Incentives for Self-Policing: Discovery, Disclosure, Correction and Prevention of Violations. December 22. *Federal Register* 60: 66705.

———. 1999. Evaluation of and Proposed Revisions to Audit Policy. May 17. *Federal Register* 64: 26745.

9

EMSs and Tiered Regulation: Getting the Deal Right

Jerry Speir

Why should the general public care about environmental management systems (EMSs)? In short, EMSs may transform the way that government regulates for environmental protection. About one-half of the states are presently engaged in environmental regulatory innovation programs (many of them built around EMSs), and the U.S. Environmental Protection Agency (EPA) recently launched the National Environmental Performance Track program (which rewards companies that implement an EMS).

For policy purposes, two essential ideas motivate most regulatory reform programs. First, the present regulatory system does nothing to encourage performance beyond the legal minimums. Second, the present regulatory regime simply fails to regulate a whole host of substances and activities with potential for detrimental effects on the environment.

The legally mandated minimum requirements of our present system, though complex, are limited. They manifest themselves primarily as (a) numbers in permits that establish limits for specific pollutants and (b) procedural requirements for monitoring pollutant discharges, reporting on that monitoring, labeling and tracking substances defined as "hazardous," and so forth. The most obvious examples of the shortcomings of the present system are the lack of restraints on energy usage, water consumption, and solid waste production. Only cost regulates how much energy an organization (or a household, for that matter) uses. It is undeniable, of course, that energy production has environmental impacts and that production is a function of demand. Although water consumption may be regulated to a degree in areas of scarcity, in most areas, cost is the principal regulator—even though excessive water usage may well have detrimental long-term environmental impacts, and even in areas where it may now seem abundant. Any organization can produce as much garbage (or solid waste) as it is willing to pay someone to haul away. Furthermore, the regulatory system focuses heavily on point sources, whereas much pollution from nonpoint sources escapes regulation.

Most substances are regulated according to their physical properties (ignitability, corrosivity, reactivity, or toxicity) or because they appear on lists of specifically regulated substances.[1] Other substances that may be harmful to the environment or public health escape regulation because they do not possess the physical properties that would subject them to regulation or because they are not presently listed. Substances that have fallen into this category in recent times include greenhouse gases, thought to be precursors of climate change; chlorofluorocarbons, responsible for atmospheric ozone depletion; and certain substances now thought to be endocrine disrupters.

A comprehensive program that encourages progressive reductions in pollution across the full range of emissions, discharges, and waste disposal practices clearly would be superior. Obviously, such a program must be voluntary. Although new regulations may still be needed in certain areas, no one realistically expects the current regulatory structure to be expanded to cover every imaginable substance and circumstance. EMSs hold the promise of helping to build such a program that might encourage the voluntary progress necessary to drive overall pollution loads downward. They offer a framework that is potentially more comprehensive than the existing regulatory structure, and they encourage continual improvement (although there is certainly an economic limitation to that impetus).

EMSs reflect a faith in process; the presumption is that a more efficient process should produce a better outcome. The theory is solid, but the proof is lacking. Critics worry that incorporating such a voluntary system into the regulatory process may risk backsliding in an organization's performance or collusion between the regulator and the regulated. Trust among stakeholders is low. Any experimentation with voluntary systems must maintain a strong option to enforce existing regulatory performance minimums and must show clear performance outcomes that are superior to the status quo. Otherwise, the credibility of the enterprise will be instantly undermined.

One might argue that the theory behind current regulatory experimentation is that people (and companies) have to be bribed. There was a joke in law school: "you regulate industry, you bribe farmers"—a simplistic summary of our historic numerical approach to putting limitations on industry on one hand and paying subsidies to farmers (to improve their stewardship) on the other. These subsidies are bribery in the simple sense of "something that serves to induce or influence" but without any immoral or illegal sense. We all like to be rewarded for good behavior. The dominant theory also hypothesizes that rewarding excellence will encourage even greater excellence as well as imitation of that excellence.

The combination of these bribery and excellence ideas leads us to the notion of tiers of regulation, which provide different treatment for high performers and laggards. First, we need a method for distinguishing those performers that are worthy of reward. Then, we need a scheme for deciding which rewards fit which behavior—that is, for "making the deal"—recognizing that these decisions are inherently subjective and not something we can turn over to the scientists in anticipation of a numerical answer. Finally, we will need a system to verify or enforce this deal. Ideally, we also would like to have a way to measure that the result of all this experimentation is superior to what went before. But lacking a reliable metric for that judgement, a credible policy decision will require consensus building. That is the role of stakeholder involvement.

Much of the present regulatory experimentation at the state level (and in EPA's Performance Track) uses this concept of different tiers. In their simplest form, these programs comprise two tiers or levels of treatment of the regulated community: one that is the existing scheme, and one that offers incentives for levels of performance higher than required. More complex programs, like Oregon's, may offer as many as four tiers, with progressively richer incentives for progressively superior performance. Because EMSs are touted as tools for improving performance, they have come to be part of the basic criteria for entry into the majority (though not all) of these tiered programs.

In light of this experimentation, I seek first to provide some background and then to answer three questions:

- Do EMSs work? Do they produce anything we might call "superior environmental performance"?
- Are EMSs essential to tiered regulatory programs that seek to differentiate between the better and worse performers in the regulated community?
- What do these tiered programs suggest about the future of environmental regulation?

As to the first question, other contributors to this volume have already answered it: not necessarily. The connection between EMS implementation and performance improvement, as Nash and Ehrenfeld's research shows (Chapter 3), is not intrinsic. The goals established by an organization in the creation of its EMS may be trivial; "form and practice may be inconsistent."

The Evidence for EMSs

We must look critically at reports of superior performance associated with EMSs. Anecdotal evidence suggests that where EMSs have been credited with verifiable improvements in performance, the real motivator often was a particular champion within the organization; much remains to be learned about whether the EMSs will survive and how they will perform when the champion departs or even when ordinary personnel changes occur.

Money is another important factor. In Nash and Ehrenfeld's study (Chapter 3), one company (Robbins) pursued ambitious pollution reduction goals, which happened to be profitable as well; another (Polaroid) modified its ambitious goals when they ceased to be profitable. And, as Nash and Ehrenfeld observe, every manager faces two separate choices: first, whether to implement an EMS, and second, how much to invest in the commitment to improvement that an EMS entails. The bottom line is the bottom line. Ultimately, investments that are positive for the environment stop happening at the point at which they are deemed too expensive, though that determination may be more complicated than a straightforward cost–benefit analysis would indicate.

Nonetheless, anecdotal evidence increasingly suggests that EMS implementation often results in measurable improvements. This evidence takes many forms. At the facility level, where EMSs are typically implemented, reports of waste minimization and cost savings are common. In Oregon, Louisiana-Pacific Corporation turned a waste stream that had been costing $100,000 a year for disposal into a $100,000-a-year product as a result of the systematic rethinking of its production process that implementation of the EMS fostered. Previously, like most organizations, the facility had focused primarily on complying with its permit limits at the end of the pipe. In the same state program, OKI Semiconductor documented a net savings of more than $35,000 per year from various process changes and a negotiated reduction of $4,000 per year in its insurance premium.

Reports of EMS impacts are frequently cast simply in terms of dollars saved. Coca-Cola claims $6 million in savings from environmental management activities implemented throughout the company (GEMI 1998). Such savings can reflect many causes, from a switch in raw materials to waste reduction to paperwork modifications. The city of Scottsdale, Arizona, for example, saved $16,000 per year simply by consolidating permits among several units as a result of its EMS implementation process. Distinguishing these kinds of savings that have little or no direct impact on the environment from those that do complicates the analysis of such cost-savings data.

But the data mount. For example, a San Francisco law firm surveyed EMSs in which it had been involved at 33 industrial installations in Europe, North America, and Asia (Pillsbury Madison & Sutro 2000). It reported

- reductions of 30% in compliance costs,
- regulatory innovation (some form of efficiency increase in reporting, monitoring, data collection, or permitting) at more than 60% of firms,
- compliance improvements (77% average reduction in compliance issues identified in internal audits), and

■ product improvements (incorporation of environmentally favorable attributes).

Although one sees in such data a predictable bias toward financial savings and regulatory innovation (which results in financial savings), compliance improvements and product improvements suggest real environmental gains; however, more analysis would be required to tease out the details.

Mercedes Benz U.S. International, Inc., a Pillsbury Madison & Sutro client, reported significant improvements in its environmental impacts from implementation of its EMS, for example, a 6.5% reduction in volatile organic compound (VOC) emissions per vehicle, a 21.5% reduction in water used per vehicle, and a 25.7% improvement in its hazardous waste recycled per vehicle (Pillsbury Madison & Sutro 2000).

The U.S. Postal Service reports several positive outcomes: increased recycling (one million tons a year, resulting in $8 million in revenue); the replacement of 15,000 exit signs nationwide with more efficient ones; and the use of alternative fuels, re-refined oil, and retreaded tires (Parry 2000).

The Irish Environmental Protection Agency requires EMSs as part of its Integrated Pollution Control Licence and reports the following improvements among its various licensees: ammonia usage reduced by 90%, methylene chloride eliminated, effluent discharge reduced by 80%, and packaging waste reduced by 50% (Larkin 1998).

In a Swedish study, almost 200 Swedish companies certified to ISO 14001 or the European Union's Eco-Management and Audit Scheme (EMAS) were asked the extent to which the companies' environmental impacts were reduced as a result of their EMSs. The vast majority of respondents indicated at least some perceived connection between EMSs and performance (Enroth and Zackrisson 2000). Interestingly, this study also found that "approximately 30% of the companies claimed that they were able to demonstrate increased revenues as a result of their environmental management work."

Although many questions remain (for example, how much of this activity would have happened anyway in a business-as-usual setting?), such reports fuel the idea that public policy should encourage EMS implementation. Several states have embarked on pilot programs to further test the notion that EMSs result in environmental gains and, as others have reported in this volume, a major data collection and analysis project is under way at the University of North Carolina at Chapel Hill (see Chapter 2).

State Innovation Programs: The Many Faces of Rewards and Benefits

The state programs vary widely. Some are purely recognition programs; others offer an array of deals to encourage superior performance. Incentives fall into half a dozen broad categories: recognition, technical assistance, financial stimuli, regulatory flexibility, relationship changes, and enforcement discretion.

Recognition itself is often denigrated as a motivator for behavioral change in the context of these programs,[2] but my own research with facility-level managers suggests that recognition may sometimes play a significant role in encouraging investment in these programs. People who believe that punishment and profit (or fear and greed) are the only real motivators should consider that recognition itself may, in some circumstances, translate into market advantage. In other cases, it may simply appeal to a manager's public consciousness (or ego) to have the facility appear as a leader during his or her watch.

In the context of such programs, recognition may be a simple certificate or plaque announced with a press release, a special flag or logo for display at the facility or on its letterhead, or membership in a special council whose members are touted as being among the state's leaders (and who doubtless have a bit more access to their state regulators).

Technical assistance also gets short shrift frequently as a motivator for change, but it too can have at least an indirect impact on a firm's bottom line. In effect, technical assistance amounts to government agencies' offering free consulting services to regulated enterprises; this approach has proven especially effective in improving the performance of smaller enterprises. The Massachusetts Department of Environmental Protection, for example, has shown significant environmental improvements with direct technical assistance to small dry cleaners, photo processors, and printers (Massachusetts Department of Environmental Protection 1998). EPA's Region 9 also has had significant success with targeted assistance to small electroplating businesses, which can have a relatively high environmental impact for their size (EPA Region 9 not dated).

Undeniably, money is a great motivator, and several of the state programs offer direct financial benefits. They range from low interest loans to tax incentives, rebates on fees, and outright grants.

The incentive in these programs that most often generates debate is flexibility in the regulatory process, some alteration of the rules by which facilities are regulated. In part, this flexibility has been created in response to widespread complaints that the existing regulatory regime is unnecessarily cumbersome, duplicative, and inefficient. Because the present system focuses on permits for air emissions, water discharges, and waste handling, the flexibility of these new state programs also tends to focus on changes to permit requirements. At least, such has been the reality to date in the United States. However, the real promise of performance-driven systems is in areas where the current permitting regime does not reach. So, it is a bit disappointing that in these state innovation programs, the deals have focused largely on changes in the permitting process.[3]

To date, discussions of the deal often start with talk of expediting, extending, or consolidating permits. For the computer industry, for example, which may change production processes once or twice a year, delays for permit approvals can be costly, and some firms are willing to commit to higher levels of performance or greater transparency in their operations in exchange for preferential treatment in the permitting process. The flexibility of the programs also may include changes to the monitoring, reporting, and inspection requirements of existing permits and regulations or changes to exemptions from specific requirements. One Oregon computer chip fabricator, for example, seeks an exemption from a particular reporting requirement of the Resource Conservation and Recovery Act (concerning monitoring for leaks from a hazardous materials storage system) on the grounds that the firm employs technology that is more innovative than the regulation contemplates and that the required monitoring and reporting is redundant and unnecessary. In fact, the equipment in question would pass muster under the more rigorous requirements of the Clean Air Act, if they applied, but the unit is not large enough to be regulated under the Clean Air Act.

At its most sweeping, the flexibility contemplated by these innovation programs involves the granting of facility-wide permits (or plantwide applicability limits [PALs], as they are often called). Most frequently, firms seeking facility-wide permits are primarily interested in developing an emissions bubble for their facility to avoid the separate permitting of each air emission point source, as the present system generally requires; they seek flexibility within the bubble (or under PALs) to trade emissions, that is, to increase emissions from one unit while decreasing them in another without having to go through a complete re-permitting process. At their most extreme, such flexibility provisions may allow the trading of pollutant discharges between media—allowing increased water discharges in exchange for diminished air emissions, for example—although proposals for these kinds of trades are very rare in these state programs. They are rare partly because of the lack of criteria for judging the merits of such trades and partly because they almost necessarily would require EPA's consent and complicate the approval process or drive the process toward the federal Project XL.

That issue of state–EPA interaction is like the gorilla in the closet of these state programs. Both Wisconsin and Oregon have elected to negotiate memoranda of agreement (MOA) with EPA before entering into regulatory flexibility arrangements with any of their organizations. Wisconsin spent more than a year and a half in the process of developing the MOA, and Oregon had a similar experience. But, even when finalized, the MOAs only really establish a process by which detailed negotiations and approvals will take place as issues arise. The phrase "federal requirement" is key to triggering that process. The issue of state–EPA relations in the context of these regulatory reform programs is huge and deserves extended separate attention.[4]

Two more incentives for participation in these experimental programs bear scrutiny here. The first is often described by the agencies as an offer of a single point of contact for the firm with the agency (that is, a single agency staff person who serves as a kind of ombudsman for the regulated firm in dealing with the agency); to the firms, this incentive often represents a change in their relationship with the agency. Under the present media-oriented regulatory scheme, regulated entities typically deal with separate agency personnel for their air issues, water issues, and waste issues—and likely with different personnel if they violate the provisions of a permit and become subject to enforcement actions.

Several of the state programs offer such a single point of contact. In my interviews in Oregon, where this process is perhaps best developed, several firms described this feature of the state program as the most important. They felt that they were dealing with a single person who understood the full spectrum of their regulatory issues and who could help facilitate their interactions with the agency. Although this kind of personalized service represents a fairly high transaction cost for the agency, it translates to time and money saved for the firms. Of course, it also raises questions, especially in the minds of public interest nongovernmental organizations (NGOs), about the independence of this single agency representative and about the potential for his or her capture by the regulated entity.

The last category of program incentives is enforcement discretion. Especially in those programs that require an EMS, one would reasonably expect that conscientious EMS implementation would reveal some problems that previously had gone unnoticed. Threatening vigorous enforcement action for every problem so discovered would provide a significant disincentive for participation in the program; however, a blanket amnesty for all such discoveries could open the door to substantial mischief. Therefore, several states—especially those that do not already have a policy covering this situation—include such a policy in their state innovation programs. Typically, these policies are modeled on EPA's Self-Policing Policy (U.S. EPA 2000) and provide various levels of amnesty, depending on the severity of the problem and the speed with which it is rectified.

For entry into these programs, states typically require some kind of performance measure that indicates that an organization is worthy of preferential treatment (often expressed in terms of a number of years without a significant violation). An EMS may or may not be required. When an EMS is required, some programs have additional requirements to account for the perceived shortcomings of most EMSs (particularly ISO 14001) as public policy instruments. These additional requirements typically address the organization's compliance with the law, openness with the public, and reporting of performance information.

State programs may or may not require beyond-compliance performance levels, and this situation has been the subject of a basic division among proponents. In some states, it is enough for innovative programs to produce the same level of environmental protection (assuming that level is at least equal to legal compliance) at a reduced cost to the regulated entity.[5] For others, only beyond-compliance performance levels are worthy of the program's benefits and the additional public resources necessary to manage the programs.

In addition to the issue of compliance versus beyond compliance, proponents divide over whether innovation programs should be open only to entities that are already superior performers or to anyone willing to commit to the program's goals (compliance or beyond compliance). For some proponents, only superior performers should reap the programs' rewards. Others, arguing that the greatest opportunities for performance gains are among the firms that are presently the poorer performers, believe that innovation programs should be open to all.

Reporting requirements among the programs vary from a requirement to produce an annual report to a more general requirement for public communication. The requirements for public participation and stakeholder involvement can range from hosting an open house to including outside stakeholders in the private firm's decisionmaking process. These public information and transparency requirements are intended to help verify and lend credibility to the individual deals that are at the heart of these programs.

State Program Comparisons

The relative importance of these various provisions is perhaps best appreciated through a comparison of several state programs.

Florida

On paper, Florida's Ecosystem Management Agreements program is farthest afield from the norm in that it requires that participants "meet all applicable standards and criteria *so that there is a net ecosystem benefit*" (Florida Statutes, ch403.0752, Pollution Prevention Act of 1991; http://www.flppr. org/law/flp2.htm#_403.0752). The program is new and only being tested in one district within the state, so it is too early to evaluate it. Reports suggest that a major benefit of the program to date has been its ability to promote communication and cooperation across jurisdictional lines, for example, among local water districts, zoning districts, and the state environmental agency. In one case, Florida claims to have created 380 acres of wildlife habitat and a mile-wide wildlife corridor, restored lake shoreline, and improved water quality.

One advantage of such an ecosystem or watershed approach to regulation, obviously, is that it promotes thinking in a context larger than a single facility. However, it also broadens and complicates the question of which incentives are appropriate for which outcomes.

Indiana

Indiana's 100% Club program takes a markedly different approach. To its "Leaders," it offers permit fee rebates, expedited permits, and a "possible reduction in inspections" (Indiana Department of Environmental Management 2000). Its criterion for membership is "full compliance with applicable environmental requirements and standards." In part, this requirement is interesting because many regulators and regulated entities alike will argue that no organization is ever in 100 % compliance; the law is simply too complex. In Indiana's case, however, full compliance is defined as meeting two criteria (Indiana Department of Environmental Management 2000):

- neither the state nor EPA "has a recent or open enforcement action against the facility" and
- "any other federal, state, or local regulatory agency with environmental, hazardous material, health, or safety authority does not have serious concerns about the facility."

So, full compliance does not mean that an organization is necessarily meeting every individual requirement of the law (although it may be). Rather, aside from this novel attention to the serious concerns of other agencies, in this context, *full compliance* means only that neither the state nor EPA has taken enforcement action against the organization. Critics would argue that it may mean only that they haven't been caught.

Interestingly, certification to ISO 14001 or an equivalent EMS is sufficient for being designated a Leader in Indiana. That criterion raises the question of the value of ISO 14001 (or other EMSs) as an adequate predictor of the superior performance that the program presumably seeks to reward. Unlike programs that treat superior performance as an issue for public information and public involvement, Indiana's program simply accepts ISO 14001 or an equivalent as sufficient evidence of superiority.

Michigan

Like Indiana, Michigan's Clean Corporate Citizen program also accepts certification to ISO 14001 as sufficient to meet its EMS requirements. No additional public reporting or stakeholder involvement is necessary. Michigan Department of Environmental Management (2000) also requires only compliance (not beyond compliance) and defines *compliance* as

▓ a lack of "significant violations";
▓ an appropriate response to any cited violations (that are less than "significant"); and
▓ a statement by a responsible official that, to the best of his or her knowledge, the organization is in compliance.

Participants in the program benefit from various kinds of flexibility in requirements for monitoring, reporting, and permitting.

Georgia

Georgia's Pollution Prevention Partners program is evidence that tiered regulatory innovation programs need not necessarily include an EMS. The focus of the program is on pollution prevention activities, and it provides variable recognition and fee reductions at three different levels of achievement, from a commitment to develop a pollution prevention program to "substantial progress" toward meeting reduction goals under an established pollution prevention program (Georgia Department of Natural Resources not dated).

Texas

Until recently, Texas boasted more than 180 participants in its Clean Industries program. (The program is being transformed into a new Clean Texas program, the specifics of which have not yet been announced.) An "internal environmental management program to assure high levels of environmental compliance" was required for entry, but the program did not specifically require an EMS, nor was "high level of ... compliance" defined. Texas added the substantive criteria of a 50% reduction in Toxic Release Inventory (TRI) emissions from a 1987 baseline. It included both a "citizens communication" component and a "community environmental projects" component, but the former could be satisfied by hosting an annual open house and the latter by helping a local school establish a recycling program (Texas 2000). Such relatively low hurdles for participation raise questions about the program's ability to truly differentiate superior performers. There is always a tension between keeping the entry-level bar low enough to attract participants, yet high enough to identify those worthy of recognition.

New Jersey

Some states deal with this problem by creating multiple-tiered programs. New Jersey, for example, has a new Silver and Gold Track program that begins with a recognition-only tier (Silver) and progresses to a flexibility tier (Gold), with an intervening Silver II Track that addresses specific greenhouse gas issues. An EMS, an outreach plan, and a "good environmental record" are the essential criteria for entry into the program. Although the Gold Track is still in development, the agency is considering, among other things, requiring: commitment to declining facility-wide emission and discharge caps; advanced pollution prevention, source reduction, recycling, and conservation of water and energy; comprehensive facility monitoring and consolidated reporting; and enhanced community outreach, as well as "in-depth process-level materials accounting" (New Jersey Department of Environmental Protection 2000). New Jersey is seeking statewide XL status for the program under EPA's Project XL.

Oregon

Oregon has arguably the most complex—and the most fully developed—multi-tiered program.[6] At its lowest Participant level, it requires (a) that the organization has "developed a program that will achieve environmental results that are significantly better than otherwise required by law ...," (b) a baseline performance report, and (c) a stakeholder involvement plan (Oregon Administrative Rules [OAR] 340-014-0115). The intermediate Achiever level requires (a) an implemented EMS (ISO 14001 or equivalent); (b) demonstrated reduction in overall environmental impacts over the previous three years; and (c) a stakeholder involvement plan that "encourages public inquiries," provides mechanisms for discussion of the firm's performance objectives and targets, and "considers the results of stakeholder involvement in [the firm's] decision-making" (OAR 340-014-0120). At its highest Leader level, the program focuses on (a) a demonstration by the firm of industry leadership in applying sustainable development principles to the environmental life cycle aspects of the firm's activities, products, and services and (b) a stakeholder involvement process that "makes efforts to establish and maintain understanding, constructive dialogue and partnership with significant stakeholders" (OAR 340-014-0125). The ambitions of this program explain, in part, why Oregon was only issued its first two Green Permits in December 2000, even though the program has been in development for five years. Four other Green Permits are under consideration.

Wisconsin

Even at its lowest level, Oregon's program is for superior performers only—those willing to perform significantly better than required by law. By contrast, Wisconsin makes possible the participation of entities that may not already be superior performers. For entry, Wisconsin requires only "at least the same level of protection of public health and the environment as current law" (Wisconsin Statutes Annotated 299.80). However, it does include significant public participation and public information requirements that might reasonably be expected to drive performance to levels beyond compliance minimums. Despite such expectations, there is no hard evidence to date. Although the program was launched in 1997, it had yet to sign its first cooperative agreement as of January 2001.

Significantly, though not surprisingly, programs with the more strenuous requirements—especially for public reporting and public involvement (as in Wisconsin and Oregon)—are finding it hard to recruit participants. On the other hand, programs with relatively low hurdles (as in Texas, Michigan, and Indiana) attract far more participants.

Arizona, North Carolina, and Colorado

Other differences are highlighted by a comparison of recent legislation in Arizona and North Carolina—differences that derive from the programs' different origins. The Arizona statute was a product of state government; the proposed North Carolina statute was introduced on behalf of an industry group.

First, as to the beyond-compliance issue, the North Carolina proposal pointedly required only "compliance ... equal to or better than ... that required by applicable environmental regulations" and specifically notes that an applicant "may comply ... by demonstrating an innovative approach *or* cost effective results" [emphasis added] (North Carolina General Assembly, Environmental Excellence Program Agreements, House Bill 1580, §113A-248(2), 2000). Arizona's statute, on the other hand, simply avoids addressing the issue, and one may reasonably surmise that proponents of the statute found the "beyond compliance" language politically untenable (especially because much of the language of the statute was borrowed from Colorado's Environmental Leadership Act (Colo. Rev. Stat. §25-6.7, 1998; http://www.state.co.us/gov_dir/leg_dir/olls/sl1998/sl.235.htm), where "beyond compliance" is mentioned prominently). Critics of the North Carolina proposal, and others like it, argue that the statute would have provided regulatory flexibility for activities that were cheaper but not necessarily connected to reduced environmental impacts.

On the issue of compliance itself, Arizona's statute has a strong requirement for documentation of "compliance with environmental statutes and rules, and other performance goals that may be contained in the agreements" (Voluntary Environmental Performance Program, Bill 1321, §49-173(B)(2), 2000). However, North Carolina's proposed statute only required "compliance with all environmental excellence and innovation goals"—not a demonstration of compliance with all existing legal requirements.

To its credit, the industry-backed plan in North Carolina required submission of a "stakeholder plan ... to identify and contact stakeholders, to advise stakeholders of the facts and nature of the project, and to enable stakeholder participation, review and comment *during the development*" of proposals

(North Carolina General Assembly, House Bill 1580, §113A-248(2), 2000).[7] A good lawyer would note that nothing in the proposed statute ever required implementation of such a stakeholder plan; only its submission was required. Such omissions only tend to heighten public interest concerns that innovation is the work of the devil.

Even such brief comparisons are instructive. First, it should be noted that not all tiered regulatory schemes have EMSs at their core. Although the basic idea of all schemes is to motivate superior performance, not all depend on EMSs as part of the assurance of that heightened performance. (Nor do they agree about the definition of *superior*.) In fact, most have an EMS component, but some (like Indiana and Michigan) accept an EMS as sufficient evidence of superiority; others (like Louisiana and others) depend on a case-by-case judgement of superiority (and may make no mention of EMSs); and still others (like Oregon and Wisconsin) have strenuous requirements for EMSs plus additional requirements. Not surprisingly, the tougher the entry requirements, the fewer the participants.

Innovation, Information, and ISO 14001

A common factor in all these experiments is that they are faced with two issues: first, developing enough information about the performance of participants to justify their differential treatment in a separate regulatory tier, and second, developing a process for deciding on an appropriate deal.

One critical question is whether an EMS is, in itself, sufficient grounds for differentiation. The answer is clear: definitely not. ISO 14001 is the EMS most frequently cited in these programs. A properly implemented ISO 14001 EMS should, in time, ensure that an organization is operating in compliance with all applicable laws. But in fact it would not be inconsistent for an organization to achieve certification to ISO 14001 and be out of compliance, because the ISO 14001 standard requires only a "commitment to compliance" for certification. This phrase is interpreted to mean that the organization has mechanisms in place that will identify noncompliance issues and bring the organization into compliance over time. The standard's requirement of a "commitment to continual improvement" may be met, theoretically, by improvement toward compliance (by an organization that starts below that level) or by trivial improvements (as Nash and Ehrenfeld point out in Chapter 3). Another more significant problem with the ISO 14001 EMS in the context of these regulatory programs is that it does not require the generation of any information of the sort needed for judging superiority.[8] State programs that make implementation of (or certification to) the ISO 14001 EMS standard (or any equivalent) sufficient for entry into their tiers are doing so blindly, on faith.

Many states recognize this problem and have taken steps to address it through supplementary reporting requirements. An EMS (ISO 14001 or otherwise) *can* produce whatever information it is asked to produce, and an EMS certainly is capable of producing the information regulators need to make informed regulatory decisions, but ISO 14001 does not make such transparency a requirement.

There is a serious question as to the kinds of information that are indeed appropriate for evaluating an organization's environmental performance, and much important work is being done in the realm of establishing useful performance measures.[9]

But even if meaningful performance data were at hand, a remaining critical issue would be, what kind of deal should be struck? What level of benefits or incentives should be matched to that performance? There is no scientific, objective answer to that question, and EMSs offer no magic bullet. It is a public and political question, and whereas some states are dealing with it in the traditional way (as the prerogative of the administrative agency)—especially in the programs with lower ambitions and entry requirements—others have cast it into the province of stakeholders and public participation. A cynic might see this trend as a clever way to defuse criticism or as a scheme for devolving the decisionmaking process (to use the jargon of a few years ago) to ever lower public levels, where expertise for meaningful critique is apt to be lacking. Proponents of this process, however, see it as a recognition of the facts that these decisions are quintessentially social (not scientific) and affect a broad spectrum of society and that the questions being decided reflect the profound uncertainties at the root of environmental law and policy.[10]

The Role of Stakeholders

Several recent studies have looked at these stakeholder issues. EPA (U.S. EPA 1999b) has written a revised Stakeholder Involvement Action Plan—one part of which is a recently completed study by the Environmental Law Institute, *Building Capacity to Participate in Environmental Protection Agency Activities: A Needs Assessment and Analysis* (ELI 1999). EPA's Common Sense Initiative (U.S. EPA 1998) has produced a report of its Stakeholder Involvement Work Group. But with rare exception, these studies address issues of public involvement in *public* processes—in agency rulemaking or Superfund cleanup decisions, for example. That matter is substantially different from the involvement of stakeholders in EMS development at a *private* facility as part of the criteria for entry into a performance tier of regulation (as contemplated in the Wisconsin and Oregon programs, for example).

This is new territory. In looking about for stakeholder involvement guides, the best I have found is EPA's *Constructive Engagement Resource*

Guide: Practical Advice for Dialogue among Facilities, Workers, Communities, and Regulators (U.S. EPA 1999a). It is a product of the Computers and Electronics Sector Subcommittee of EPA's Common Sense Initiative. In addition to offering guidelines, it includes 11 diverse and informative case studies. Most directly applicable to programs such as those in Oregon and Wisconsin are the case studies of the Intel and Lucent XL stakeholder or constructive engagement processes. (The guide defines *constructive engagement* as "any effort that brings together a diverse group to cooperatively discuss a facility's environmental activities" [U.S. EPA 1999a, 1].) Where there has been success, it has been attributed to unsurprising things: to committed individuals (at Lucent, a facility coordinator and a professional facilitator) and to openness and frankness.

An apparent turning point at Lucent "came when the plant had a minor spill," reported it to the group, and got a general reaction of, "Well, it looks like you handled things well." As one member of the group put it, "It was in effect [Lucent] blowing the whistle on themselves.... It's healthy for people to admit their mistakes, to admit they're human, and invite us to help keep them clean" (U.S. EPA 1999a, 92).

However, it should not be surprising that for some public interest observers, such anecdotes make them decidedly uneasy. They fear something akin to the "capture" problem. As one put it to me, "I worry about a [citizens' advisory panel]-type group becoming too 'understanding' of the company's problems and limitations for their own good. Someone's got to be able to stand outside and say 'I don't give a damn—it's not my problem—fix it!' without worrying about it. Who assumes that role if the local folks are inside? Or isn't there the possibility of fomenting community division between the 'ins' and the 'outs'?"

Recently, environmental groups have put forth some of their own evaluation criteria for innovation projects, criteria that have a direct bearing on the stakeholder involvement issue and, by extension, on the question of "getting the deal right." A group coming together under the banner of the Pollution Prevention Alliance (with representatives from Citizens for a Better Environment, Environmental Defense, Friends of the Earth, National Wildlife Federation, the Good Neighbor Project, U.S. Public Interest Research Group, and several others) has issued such a document, *Alternative Regulatory Pathway: Evaluation Criteria* (Pollution Prevention Alliance 1996). It lists seven components that it regards as "the building blocks for a credible ARP [Alternative Regulatory Pathway]":

1. Eligibility to participate. They see such programs as solely for leaders who are willing to "involve stakeholders from the outset, typically by forming full partnerships of public interest groups, businesses, and governments to cooperatively manage the project."

2. Clear and ambitious performance goals.
3. Enforcement and incentives to perform. They support compliance assistance but also insist on "consequences (for example, per unit fee) for short-term minor noncompliance."
4. Accountable results. They support consolidation of permits "to eliminate redundancies and better match the way businesses make decisions" but insist on a "reporting system ... understandable to all"—and a "materials accounting approach that tracks both inputs and outputs ... [as] the optimal ARP metric."
5. Public participation, including an "inclusive and meaningful decision-making process."
6. A multimedia whole-facility approach, operating from "a baseline from which ongoing performance will be evaluated."
7. A life-cycle perspective that "encourage(s) producers to reduce the life-cycle environmental impacts of their products."

More recently, a group of Wisconsin NGOs devised a 14-point test for evaluating reinvention projects.[11] First are the *why* questions: Why can't the project be done under existing law? Then come the *performance measure* questions: How will we know it's better? Are there clear goals and objectives? When and how will evaluation be done—and against what baseline? Then there are the *process* questions: Who's involved? How will decisions be made? Do citizens have a meaningful role (meaning "the potential and opportunity to affect the outcome")? Will stakeholders have the necessary resources to complete the project?

The 14 points express a concern for enforceability and costs—a concern for the trade-off between performance and regulation. What regulatory relief is being considered? What will be the corresponding public benefits?

The constructive engagement case studies and the NGO evaluation criteria suggest quite a bit about the *how* of stakeholder involvement. It will not be easy. It will be very time-consuming and resource intensive. And the highest hurdle, from the perspective of the public interest stakeholders, is apt to be in the realm of providing enough opportunity for influencing outcomes to make participation worth the effort. Serious questions remain about a state's ability to maintain such activities for a large number of projects and over an extended period of time.

But even more fundamental than the question of *how* to run stakeholder involvement processes is the question of *why*. From a public interest perspective, if stakeholder involvement processes are apt to tend more toward the advisory than the decisional (which seems the likely outcome), is the involvement process really worth the effort? Might not everyone be better off with simply an improved information system with an efficient mechanism for publicly raising questions about the importance of the information and for having those questions addressed? If that mechanism requires something like a stakeholder panel (for oversight), might we not at least reduce the potential for overload by having one such panel per state (or region of a state)—rather than citizens' advisory panels at each facility or in each community?

The most promising stakeholder processes are those in California and Oregon. Oregon managed a single statewide stakeholder involvement process for the development of its Green Permits program. California is now operating two regional stakeholder involvement groups (one for northern California and one for southern) for its EMS Innovation Initiative, which also appear to be effective (California EPA 2000).

EPA's Performance Track

The state experiments (especially those that attempt significant innovation) are not only high in transaction costs for stakeholders but also very staff-intensive. The high transaction costs are one of the principal targets of EPA's recently launched National Environmental Performance Track program, which seeks to accomplish goals similar to the state programs but tries to do it by category rather than site-by-site.

The program lists some very specific reporting and monitoring benefits, such as "reduced Clean Air Act reporting and record keeping for Maximum Available Control Technology (MACT) facilities that reduce their emissions below the threshold for major sources" for its initial Achievement Track. For these kinds of benefits, facilities must implement an EMS and commit to certain public outreach and reporting requirements. A higher Stewardship Track is promised for May 2001.

The relationship between this new federal program and the existing state programs will be important to watch. In recent interviews, officials in the state programs have been willing to give the federal program a chance, hoping that the two will be compatible, and that the state programs will still be able to accomplish more detailed innovations than those to which the federal program currently aspires. However, the same officials recognize that the federal program could subsume the state programs.

Next Steps?

Yet another tack on tiered regulation is at a fledgling stage in Wisconsin. Referred to as the Green Tier, the idea is based on contract law and borrows significantly from such European programs as the Dutch Covenant and Bavarian Compact systems. Under this scheme, not only the incentives but also the consequences of failing to live up to commitments (by both parties, the regulated entity and the state) would be negotiated.[12] The program also suggests the possibility of new judicial remedies, potentially taking regulation into the realm of contract litigation. Proponents hope the program can be focused on areas that are not heavily regulated presently (such as agriculture), and they also hope to keep its transaction costs low, but how they will do that remains to be seen.

The contract idea is an intriguing one, and the concept could bear fruit. But if it does, it will face the same issue as the current state experiments: tension between the states and the EPA over how far the states can deviate from federal requirements. There will be the same issues about information (how to generate it, how much is enough) and the same potential for EMSs to produce it (or not). And there will be the same concerns about the public process for approving the deal—in this case, the contract.

Continuing in the spirit of the states as "laboratories of democracy," Oregon has yet another relevant innovation afoot. On May 17, 2000, Governor John Kitzhaber signed an Executive Order on Sustainability (Executive Order No. EO-00-07; http://pub.das.state.or.us/sustainability/exec_order.htm), promising that "The State of Oregon shall develop and promote policies and programs that will assist Oregon to meet a goal of sustainability within one generation—by 2025." It will start, appropriately, by looking internally at practices within state government. If successful there, it almost certainly will be turned outward, with the goal of improving environmental performance and environmental protection across the regulatory landscape. Although its scope may well be broader than the current crop of state innovation programs, the order will face the same need for a public judgement about how much—pollution, regulation, and information—is enough.

Conclusions

In sum, there is pressure for change. At its best, it is motivated by a desire for improved environmental protection and superior environmental performance, recognizing that we are dealing with quintessentially uncertain matters and that compliance with the law does not address all risks. But always there is the fear—and the possibility—that these innovation programs might be masking regulatory rollback.

Differentiation is the order of the day. It is thought that the way to promote superior performance (that is, beyond-compliance performance) is to treat the good guys better than the bad guys—and to encourage those good guys to perform even better by providing incentives. The essential need is for information on which to base the differentiation and a process for deciding on the deal, followed by a process for enforcing the deal.

Lots of experiments are under way at the state level—with some kind of tiered regulatory system for distinguishing performance levels and bestowing rewards. All programs seek to promote superior performance, but some define *superior* as the same protection at cheaper costs. Some programs have public information requirements, and others do not; those requirements can vary from hosting an annual open house to involving the public in private decisionmaking processes. Many, but not all, require EMSs. A few of them make implementation of (or certification to) the ISO 14001 EMS (or an equivalent) sufficient to garner the benefits of the program.

One of the few things that can be said with certainty is that ISO 14001 is not, of itself, adequate basis for that kind of public policy decision. ISO 14001 guarantees no particular level of performance, and without supplementation, it provides no public information by which to judge an organization's performance. Still, it provides a standardized framework and, with additional elements, *can* generate that necessary information.

The process for deciding which levels of performance deserve preferential treatment is a public and political one, and EMSs provide no quick fix. In many of the state experiments, that decisionmaking process is in the province of stakeholders rather than in the traditional administrative arena, and much remains to be learned about the viability of those stakeholder processes. In a sense, all these programs reflect efforts to do more with less, and the measure of their success will be whether they can demonstrate that the outcome really is more. That demonstration depends on a system of information based on performance indicators that we are only beginning to build.

Acknowledgements

The author's initial study of the state innovation programs in Oregon and Wisconsin was part of a larger review of state–EPA relations conducted for the National Academy of Public Administration, whose support he gratefully acknowledges.

Notes

[1]See, for example, the hazardous waste lists under the Resource Conservation and Recovery Act (*Code of Federal Regulations,* Part 261, Title 40, 2000).

[2]See for example, Crow 2000 (24).

[3]This situation may change with the implementation of some projects presently under discussion, such as Wisconsin's Green Tier program, which contemplates performance contracts that would reach "unregulated aspects," to use the jargon of ISO 14001.

[4]As one observer has accurately noted, "The future path of a performance track system in the United States ... rests heavily upon the ability of states and EPA to coordinate their initiatives" (Crow 2000, 29).

[5]See, for example, Louisiana's Environmental Regulatory Innovations Program (Louisiana Administrative Code §33:I.3703), which authorizes "regulatory flexibility" for "superior environmental performance" and defines the latter in terms of lower pollution levels or the same levels achieved at lower costs.

[6]For more information, see http://www.deq.state.or.us/programs/greenpermits/gpupdate.htm.

[7]The timing of stakeholder involvement is important. Involvement of relevant stakeholders during the development stage of projects is far superior to a mere review function at the end of the process.

[8]ISO 14001 has no requirements for the provision of information to entities outside the organization, except for the information it provides to auditors. For an analysis of the shortcomings (and potential strengths) of ISO 14001 as a public policy tool, see Morrison and others 2000.

[9]See, for example, the work of the Global Reporting Initiative (http://www.globalreporting.org/index.htm) and the Environmental Compliance Consortium (http://www.complianceconsortium.org/).

[10]Heinzerling 2000 is a useful recent survey of the uncertainties issue.

[11]Irreverently called *A Sniff Test for Evaluating Environmental Regulatory Reinvention Projects*, the test was compiled by Steve Skavroneck, Liz Wessel, Caryl Terrell, and Susan Mudd, representatives of Citizens for a Better Environment and the Wisconsin Sierra Club Chapter (http://www.dep.state.pa.us/dep/deputate/pollprev/mswg/ltconf/skavroneck/index.htm).

[12]A consistent complaint in some innovation programs has been of the state's inability to live up to its commitments.

References

California EPA (Environmental Protection Agency). 2000. Environmental Management and Sustainability Program: Innovation Initiative—Environmental Management System Project. June. California Environmental Protection Agency, 5.

Crow, Michael. 2000. Beyond Experiments. *The Environmental Forum*, May/June, 19–29. (Based on a study by the Tellus Institute.)

ELI (Environmental Law Institute). 1999. *Building Capacity to Participate in Environmental Protection Agency Activities: A Needs Assessment and Analysis*. http://www.eli.org/ (accessed January 13, 2001).

Enroth, Maria, and Mats Zackrisson. 2000. Environmental Management Systems—Paper Tiger or Powerful Tool? Presented at the 2000 Eco-Management and Auditing Conference, June 29–30, 2000, University of Manchester, U.K. (on file with the author).

EPA (Environmental Protection Agency) Region 9. Not dated. *Merit Partnership Pollution Prevention (P2) Project for Metal Finishers*. http://www.epa.gov/region09/cross_pr/merit/metal.html (accessed October 16, 2000).

GEMI (Global Environmental Management Initiative). 1998. Environment and Business. http://www.gemi.org/docs/PubTools.htm (accessed January 13, 2001).

Georgia Department of Natural Resources. Not dated. *Pollution Prevention Partners.* http://www.ganet.org/dnr/p2ad/recog/p3.htm (accessed October 16, 2000).

Heinzerling, Lisa. 2000. Pragmatists and Environmentalists. *Harvard Law Review* 113: 1421–1447.

Indiana Department of Environmental Management. 2000. *100% Club Fact Sheet.* http://www.state.in.us/idem/100percentclub/factsheet.html (accessed October 16, 2000).

Larkin, Padraic. 1998. Incorporation of Environmental Management Systems into Integrated Pollution Control Licensing in Ireland. Proceedings of the Fifth International Conference on Environmental Compliance and Enforcement, November, Monterey, CA.

Massachusetts Department of Environmental Protection. 1998. *Massachusetts Environmental Results Program.* http://www.state.ma.us/dep/erppubs.htm (accessed October 16, 2000).

Michigan Department of Environmental Management. 2000. *Clean Corporate Citizen Program.* Rule 324.1507. http://www.deq.state.mi.us/ead/tasect/c3/ C3GENezread. pdf (accessed October 16, 2000).

Morrison, Jason, and others. 2000. *Managing a Better Environment: Opportunities and Obstacles for ISO 14001 in Public Policy and Commerce.* Oakland, CA: Pacific Institute.

New Jersey Department of Environmental Protection. 2000. *Silver and Gold Track Program for Environmental Excellence.* http://www.state.nj.us/dep/special/silver/ index.html (accessed January 13, 2001).

Parry, Pam. 2000. *The Bottom Line: How to Build a Business Case for ISO 14001.* Boca Raton, FL: St. Lucie Press, 62–66.

Pillsbury Madison & Sutro. 2000. EMS Survey Data. San Francisco, CA: Pillsbury Madison & Sutro. (On file with the author.)

Pollution Prevention Alliance. 1996. *Alternative Regulatory Pathway: Evaluation Criteria.* http://alt-path.com/arp.htm (accessed January 13, 2001).

Texas. 2000. *Clean Industries 2000.* http://www.newenvironmentalism.org/priv_sec/ tx-clean.html (accessed January 13, 2001).

U.S. EPA (Environmental Protection Agency). 1998. Common Sense Initiative. June. http://www.epa.gov/commonsense/CSIinfo.html (accessed January 13, 2001).

———. 1999a. *Constructive Engagement Resource Guide: Practical Advice for Dialogue Among Facilities, Workers, Communities, and Regulators.* EPA 745-B-99-008. http://www.epa.gov/stakeholders/pdf/resolve2.pdf (accessed October 16, 2000).

———. 1999b. *Stakeholder Involvement Action Plan.* August. Washington, DC: U.S. EPA.

———. 2000. Incentives for Self-Policing: Discovery, Disclosure, Correction, and Prevention of Violations. April 11. *Federal Register* 65: 19617–19627.

Conclusion

10

Toward a Management-Based Environmental Policy?

Cary Coglianese and Jennifer Nash

Private firms and public environmental agencies today are devoting increasing amounts of attention to environmental management systems (EMSs). Managers of private organizations are learning about the EMSs that their competitors and customers have adopted and are considering the strategic advantage of implementing similar systems. They are assessing the environmental impacts of their facilities, setting goals to reduce those impacts, documenting procedures, training workers, measuring progress, and engaging third parties to assess their systems. Encouraged by the results of the EMSs they have implemented, some managers are requiring that their suppliers adopt similar systems as a condition of business. Public policy makers, too, are paying close attention to EMSs.

More than a dozen states have begun to create tiered regulatory systems, making entry into the privileged tier dependent on EMS adoption. The U.S. Environmental Protection Agency (EPA) has launched the National Environmental Performance Track, which offers recognition and other incentives to facilities that adopt EMSs with certain characteristics.

What are the benefits of EMS adoption, and how should public policy adapt, if at all, to their widespread use? In this chapter, we revisit the main lessons from the research presented in this book and point out the direction for future research that will be needed to determine how far current policy should shift toward becoming a management-based system of environmental policy.

Benefits for Firms and Society

One clear lesson is that companies respond differently to environmental pressures. In many firms, the natural environment is still only a peripheral factor in business decisions. It is rarely discussed except in the context of regulatory compliance. The environmental manager's primary function is "chief compliance officer" who makes sure permits are up-to-date and control equipment is operated as specified so business managers will not have to concern themselves with the environment at all.

But for other firms, including many of the ones mentioned in this book, the environment has assumed an altogether different importance. Environmental performance is viewed as a business need. Managers are attuned to various external and internal actors who value environmental performance: customers, competitors, shareholders, insurers, environmental advocacy groups, regulators, labor unions, and employees. The managers of these firms may or may not be environmentalists themselves; they are simply good managers. If they want to build their customer base, borrow capital, buy insurance, and attract skilled workers, they must invest in improving environmental performance. For such companies, the definition of *strong environmental performance* is expanding. In the past, compliance was sufficient; now it is only a first step. The managers' job is to find ways to reduce materials inputs and waste by tightening operating processes and bringing environmental concerns into business planning. They engage community residents and other people whose values may be different from their own in an effort to identify new strategies. They work to ensure that foreign facilities that operate under far less stringent regulatory regimes meet the same standards as home-country plants.

EMSs are a part of many of such companies' environmental programs. They provide a way for managers to institutionalize corporate environmental goals and the practices that will work toward achieving those goals. The authors of this book present arguments for the benefits EMSs can provide. In Chapter 2, Andrews and others suggest that, over time, EMSs have the potential not only to improve regulatory compliance but also to create a system of continual improvements toward reducing a facility's most pressing environmental problems. Most of the facilities participating in the National Database on Environmental Management Systems (NDEMS) used the EMS design process as an opportunity to examine all their activities to identify those that would have a potential impact on the environment. EMSs have the potential to engage managers in an investigation of the root causes of environmental problems, allowing them to prevent—not only control—pollution.

Andrews and others explain that EMSs offer benefits because they operationalize management commitments. Managers may know and understand that business and environmental interests increasingly intersect but fail to act effectively. Compensation systems may not reward environmental performance, for example, and work routines may keep environmental managers organizationally separated from process engineers. EMS design is a deliberate process in which managers assess and prioritize environmental impacts and determine how best to focus organizational attention on reducing those impacts. When designing an EMS, managers have the opportunity to correct mistakes in established practices that have kept business and environmental interests from meshing. The EMS process establishes a cycle in which managers continually seek better outcomes by setting targets, establishing routines, checking progress, and striving to do better next time. People from diverse functions may take part in this process so that environmental managers are no longer lone voices urging attention to environmental needs; they are joined by marketing managers and process engineers who have come to see environmental protection as their job, too. In this way, as Coglianese (Chapter 8) observes, EMSs may "draw in" managers and employees who otherwise would be left out.

Research presented in this book suggests that facilities with EMSs may perform better than the norm in terms of several criteria. EMS adoption appears to correlate with advanced management practices generally. Florida and Davison (Chapter 4) find that managers of EMS plants who also have pollution prevention programs are likely to adopt a "broad bundle" of innovative approaches such as total quality management, employee involvement, and performance measurement systems. Managers use EMSs as one of many approaches to make their firms more competitive.

Facilities with EMSs also may pose lower environmental risk than comparable plants without such systems. Although there is little direct research yet that compares the environmental risks posed by EMS plants with those posed by non-EMS plants, Florida and Davison report that at least EMS plant managers tend to rate their environmental performance better than managers of non-EMS facilities. Managers of EMS plants are far more likely than their non-EMS peers to cite recycling, air emissions reduction, and solid waste reduction as sources of environmental performance. By their own estimation, they cause fewer environmental problems for surrounding communities.

EMSs may help managers reduce the costs of their environmental programs. Speir (Chapter 9) includes examples of firms that turned waste streams into products, decreased water and energy use, and therefore reduced their utility bills. Numerous organizations appear to reduce compliance costs after EMS adoption. Exactly how these cost savings are achieved is not yet clear. Presumably, as more workers are drawn in to help meet a facility's environmental goals, more ideas are generated about how to do so efficiently. The savings reported in some facilities are substantial. Some firms that report financial savings from EMS adoption had already implemented pollution prevention programs, so their ability to identify further cost reductions is even more notable.

The chapters in this book make the case that managers can use EMSs to achieve important benefits in terms of environmental performance and cost reduction. Most of the research presented addresses the role of EMSs in organizations in the United States. As Panayotou argues (Chapter 5), these systems may have even greater value in developing countries, where environmental regulatory systems are less established. By using a common EMS framework, managers working in multinational corporations can create a unified approach in all of their operations. By their nature, EMSs are flexible, and firms can choose their own goals and the pace at which they will move toward them. A common EMS provides opportunities for sharing ideas and practices across multiple national jurisdictions without imposing unreasonable costs.

Policy Responses

How should environmental policy respond to the widespread adoption of EMSs? We have argued that EMSs allow managers to operationalize their commitments to strong environmental performance. In some cases, firms can use EMSs to outperform their non-EMS peers, perhaps even posing lower risks to their communities and achieving greater efficiency. Many policymakers in business and government see the movement toward EMS adoption as an opportunity for large-scale changes in the regulatory system. What would a management-based environmental policy look like? Would such a system be desirable?

Conceivably, environmental policy could shift from a system that relies on technology-based standards to a system built more explicitly around performance standards. In the EMS design process, managers generate performance targets on the basis of their understanding of their company's most significant impacts. These targets could become the basis of a performance-driven regulatory system, especially if progress toward meeting the targets were regularly monitored and periodically verified by qualified third parties. By setting performance targets instead of technology standards, firms would retain flexibility in selecting the means to achieve these targets, allowing them to choose the lowest-cost method of making environmental improvements. Performance targets could be set for pollutants in any media, allowing firms to "trade" between water and air emissions. Eventually, with additional developments in risk analysis, firms' targets could be set in terms of the overall risk created by the firm. Firms that secured reliable third-party audits of their environmental performance as part of an EMS could pave the way for a potentially dramatic shift toward a much more flexible style of regulation.

With a management-based environmental policy, government agencies could more efficiently allocate their monitoring, permitting, and enforcement resources. With effective EMSs, firms would engage in what essentially amounts to a system of self-regulation, but still one with the threat of regulation or other policy incentives in the background. Government agencies could require new reporting requirements, under which firms and independent auditors would become the principal monitors of environmental performance. They could rely on the information generated by these reports as a basis for assessing compliance or allocating regulatory resources. The mere presence of a verifiable management system that included internal and third-party auditing would provide assurance that a firm's environmental impacts were being well managed. Ultimately, government agencies could shift resources toward managing a system of management systems.

Over time, the widespread use of EMSs might create changes in the relationships among businesses, environmental groups, and local communities. If the information generated by EMSs were readily accessible on the Internet, community groups could closely monitor the environmental impacts of local firms. If firms routinely used a transparent, systematic process for their environmental management, outside organizations could more feasibly provide input into that process. With transparent EMSs, these organizations also may be able to participate more effectively in government decisions about setting a firm's performance targets.

The transformed environmental regulatory system described in the preceding paragraphs offers distinct advantages: flexibility, efficiency, and transparency. If firms widely incorporated EMSs as a normal part of corporate management, government agencies might be able to entertain more performance-based environmental regulation, new methods of tracking environmental information, and more strategic approaches to regulatory enforcement. Overall, however, the authors in this book suggest some caution before accepting EMSs as a basis for public policy. They argue that whereas many managers may find EMSs an important and helpful tool, public policy designed to encourage or incorporate EMSs into a regulatory strategy may steer both firms and agencies in the wrong direction. EMSs may incorporate goals that represent managers' aspirations, but not society's. They may distract agencies from more promising policy options. The incentives agencies provide may encourage EMS adoption, but not the underlying managerial dedication that makes such systems meaningful. Furthermore, EMS-based policy initiatives may consume the resources of firms and agencies without leading to results that would justify the commitment of these resources.

Moomaw (Chapter 6) argues that the performance targets managers select for their EMSs may not correspond to society's goals. The performance targets that managers choose tend to focus narrowly on compliance and emissions reductions, whereas society's goals are much broader and include equity, fairness, and trust. EMSs have done little to address these goals in general. In fact, several companies that have gone far to act in a sustainable manner have not created formal EMSs at all. They simply respond, Moomaw argues, with new programs that address perceived needs, such as a housing program for impoverished workers.

Metzenbaum (Chapter 7) argues strongly for the benefits of performance-focused regulation but warns that EMSs are the wrong tool for moving policy toward this end. Performance goals allow organizations to experiment, learn, and innovate and can motivate progress, even when they are not explicitly linked with rewards. Yet EMSs are not the place where policymakers should focus, she argues. Some EMSs do not ensure that managers will focus on environmental outcomes and do not encourage firms to disclose information about environmental performance. The emphasis on EMSs may distract policymakers from a more pressing concern: the generation of credible and comparable environmental outcome information.

Furthermore, public policy that encourages EMSs may fail to foster the underlying motivations needed to make them work. Although research presented in this book suggests that EMSs can lead to environmental improvements, researchers have yet to explain why they observe firms realizing these gains. Coglianese (Chapter 8) raises the question of whether these results are due to the EMS or to something else about the adopting organization. Florida and Davison (Chapter 4) show that firms that adopt EMSs tend to be large organizations that have already adopted advanced management practices. Firms that use EMSs may well be different from the typical firm. The effectiveness of their EMSs may depend on their special resources, capabilities, and motivations. The distinction between whether an EMS itself or the characteristics of the adopting organization bring about improvement has important implications for policy. If the effectiveness of the EMS depends on something other than the system itself, public policies that attempt to foster EMS adoption will likely fail to promote environmental protection. Conceivably, such policies could even weaken the motivation of management to seek environmental improvement.

Finally, the presence of an EMS, particularly one based on ISO 14001, is not necessarily a good metric for differentiating among firms. Nash and Ehrenfeld (Chapter 3) and Speir (Chapter 9) make this argument most clearly. Firms can adopt an EMS without investing in environmental performance improvement. Only through careful observation of a firm's environmental targets and performance over time can agencies assess whether a firm's EMS is intended to build the legitimacy of the organization or designed to motivate and guide action.

Furthermore, meshing performance-based programs with traditional regulatory programs will be difficult, as several authors point out. In theory, at least, agencies could offer incentives powerful enough to change the values of managers. They could offer firms with particularly strong EMSs exemption from regulatory requirements that are particularly costly, or they could decide not to subject these facilities to regular inspections. In practice, however, agencies are constrained when it comes to providing meaningful benefits. Agencies face a trade-off between giving more and asking more. That is, the more government seeks to give managers by way of an incentive, the more it will ask for in terms of proof that those managers' firms are deserving. This is why most performance-based programs, such as EPA's Project XL, have attracted relatively few participants. When the benefits are meaningful, the costs of participation are too high for most organizations. EPA's National Environmental Performance Track so far has attracted a comparatively large number of participants, because the costs of admission are quite low. The benefits to firms, in terms of substantive reforms to the regulatory system, are correspondingly small.

The conclusion we draw from the chapters in this book is that although EMSs may be an effective tool that managers can use to achieve their environmental objectives, the best policy response to their widespread adoption may be no response at all. It is, after all, quite possible for the widespread use of EMSs to coexist with the current system of environmental regulation. Furthermore, private mandates may be far stronger than public ones in encouraging EMS adoption. When a customer announces that it will require its suppliers to register to ISO 14001 as a condition of business, firms that rely on this customer for a substantial portion of their sales will face an overwhelming incentive to adopt that management system. Certainly, such an incentive is more likely to get the attention of managers than the package of benefits being offered in agency performance track programs.

New and Remaining Questions

The research presented in this book provides insight into the potential value of EMSs as well as their implications for public policy. Yet for as many questions that the preceding chapters have answered, many new and unanswered questions remain. A major goal of this book has been to set the future agenda for research on EMSs.

Some of the research presented here suggests a correlation between EMS adoption and strong environmental performance. However, existing research cannot yet discern whether the implementation of an EMS is itself a necessary or sufficient condition for real environmental improvement. Much of what we currently know about EMSs has been drawn from close study of organizations that have strong environmental programs. Researchers understand less well how the managers of more typical firms—or even firms that are atypical in their disregard for the environment—use EMSs.

At the heart of the research presented in this book is a question of causality. Does implementation of an EMS cause environmental performance to improve? The conclusion we draw from the preceding chapters is equivocal: it depends. However, the research presented does mark an advance by moving beyond anecdotal observations about EMSs and beginning to provide more systematic consideration of their impacts. Anecdotal observations have their place; they may inspire a manager, regulator, or scholar to take notice of a promising new phenomenon. But anecdotes can never establish that a causal relationship exists because they cannot rule out the numerous rival hypotheses that could explain improvements after EMS implementation.

Causality cannot be directly observed and can be inferred only through a carefully designed research inquiry. The classic experimental research design for assessing the impact of a particular policy tool (such as an EMS) would start by randomly assigning individuals to one of two groups: an experimental group and a comparison group. These two groups would be equivalent in many ways but would differ in others, particularly in that the experimental group is subject to a "treatment," the innovation that is to be tested. In an idealized setting, we would randomly assign facilities in our experiment group to be "treated" with an EMS. We would then observe environmental performance in both groups on two occasions, before the EMS was introduced and after it had been in place for an appropriate period. The difference in the environmental performance of the two groups at similar points in time would indicate the effectiveness of the EMS. If we did our work well, we would be able to generalize from this experiment to draw conclusions about the effectiveness of EMSs in organizations at large.

The challenge for future research will be to establish research strategies that approximate this ideal research design. It will be important to assess the extent to which EMS implementation leads to environmental and efficiency improvements by comparing EMS firms with firms that have yet to adopt formal EMSs. Far too often, attention in policy circles has focused on only firms that have adopted EMSs. To establish a causal claim that EMS adoption leads to environmental or efficiency gains, it will be necessary to compare systematically the outcomes achieved by firms with and without EMSs and to try to control for other factors that affect organizational performance.

In making such a comparison, another difficulty arises, because EMSs represent a wide range of approaches. The definition we have used throughout this book—a set of rules and resources managers adopt to develop routines that help an organization achieve corporate environmental goals—is broad. EMSs can take many forms and are as different as the many organizations that implement them. Indeed, flexibility is a chief appealing feature. EMSs allow managers to choose for themselves the impacts they consider most important and the level of resources they will provide.

Although flexibility may make EMSs attractive tools for facility managers, it presents a problem for people seeking to understand their effectiveness. The treatment we are attempting to understand is in practice many different treatments, each designed to fit the needs of the particular organization. Some managers may adopt EMSs with trivial goals and empty commitments and fail to inform workers of the system. Others will choose rigorous goals and seriously pursue them. Both kinds of approaches—and everything in between—are called EMSs. Researchers must consider how these differences might affect their research results, and policymakers should consider these differences before selecting strategies that rely on generic tests of EMS adoption.

The NDEMS project (Chapter 2) provides a rich source of data on the different ways EMSs are developed and used. This research will be especially useful to policymakers and researchers who are developing hypotheses about which EMS attributes may be most important in producing particular outcomes. For example, we will need to consider whether strong environmental reporting makes a difference in terms of performance. We also might wonder whether EMSs (and EMS standards) that specifically call for improvements in environmental performance lead to improved outcomes over those that require only improvements in the management system itself. These kinds of questions will be productively raised by the NDEMS project, which documents variation in firms' implementation of EMSs.

In addition to variable implementation of EMSs, another important issue facing future research will be selection bias. When comparing the performance of different firms, it will be important to acknowledge that the type of firm that adopts an EMS may well be different in relevant ways from the type of firm that does not adopt an EMS. Absent a mandate, managers decide for themselves whether to adopt an EMS and what level of resources to apply. Given this fact, researchers will need to untangle the effects of the EMS independent from the effects of other organizational variables, including those related to the decision to adopt the EMS in the first place. This concern is emphasized in Chapter 1 as well as in Chapters 3 and 8, but the point bears repeating.

A related issue is that environmental performance may be affected by the interaction of different factors. Florida and Davison's research (Chapter 4) is notable for the relatively large number of firms and the diversity of types included in the survey sample, enabling the researchers to compare environmentally advanced plants with nonadvanced plants. The "treatment" they studied, significantly, is a combination of two factors: EMS adoption and a history of commitment to pollution prevention. Although combining these two attributes makes it more difficult to understand the role of EMSs alone, in fact, most policy proposals to establish tiered regulatory systems do not rely on EMS adoption as the sole criterion for entry into the above-average group. They often require investments in pollution prevention as well. Moreover, managers who adopt advanced practices in one area are likely to be the kind of managers who adopt advanced practices in other areas as well. Florida and Davison put forth a plausible case that firms with EMSs and pollution prevention programs tend to pose less risk to the communities in which they are located. Their study raises an important issue for future research: namely, how to identify whether specific advanced practices, such as an EMS by itself, significantly affect environmental performance or whether performance depends on the interaction of more than one advanced practice and even other organizational characteristics.

In seeking to correlate environmental management practices with improved environmental performance, EMS researchers should search for direct ways to operationalize environmental performance. Environmental performance can be measured in many ways. The studies reported in Chapters 2 and 4 rely on the reported perception of the environmental manager as the indicator of environmental performance. The environmental manager is perhaps the person most knowledgeable about a facility's environmental performance, understanding its sources of risk, the degree to which the organization is addressing them, and how current performance compares with past practice. Yet relying on the perceptions of facility environmental managers probably also biases research results, because most managers presumably tend to view performance as having been improved under their leadership. Future research should, where feasible, use external performance measures such as environmental permit conditions, effluent discharge rates to wastewater treatment facilities, toxic releases, and environmental technology used. These measures, although sometimes difficult to gather and possessing their own limitations, would make an important advance over reliance on managers' subjective assessments.

In addition to investigating the impact of EMSs on environmental performance, the chapters in this book raise important research questions about the impact of public policy. A key set of questions centers on the degree to which existing state and federal environmental regulatory systems serve as a barrier to (or a driver of) EMS adoption, as well as how new policies might better encourage improved environmental management. Do the new incentives being offered by agencies have the potential to change the behavior of firms and draw more organizations into the class of firms considered excellent? Or do such incentives mainly attract firms that are already strong performers? What kinds of incentives might bring about innovation? In addition, we should consider how various policy changes would affect the behavior of agencies. Will new performance track programs focus agency resources on monitoring the environmental practices of firms whose performance is already strong, or will they allow agencies to shift resources and attention to poorer performers? What limited evidence exists about the experience of these programs suggests the answer may be the former.

We noted earlier that many EMSs explicitly complement regulatory programs by establishing compliance as an EMS goal. As discussed in Chapter 1, Louisiana-Pacific's overriding need to meet compliance obligations drove the firm to adopt an EMS and to raise the priority given to environmental protection throughout the organization. Indeed, a primary motivation for adoption cited by managers (discussed in Chapters 2 and 4) is compliance assurance. Because evidence suggests that the need to comply with government regulation drives EMS adoption, what should be the response of government? To foster more widespread adoption of EMSs, legislatures and agencies could conceivably increase the stringency of regulatory requirements, not move in the opposite direction.

Future Research Strategies

Public policy generally aims to address all firms, not only those that are doing well already. Future research strategies should focus on how EMS adoption shapes the environmental performance of typical facilities as well as environmental leaders. One potentially promising approach would be to take advantage of the natural experiment of customer EMS mandates. Large firms are increasingly requiring their suppliers to adopt EMSs, often modeled on ISO 14001, as a condition for business. EMS mandates usually apply to all suppliers of a certain size or type, regardless of the strength or weakness of their environmental programs. By examining EMS adoption in these diverse organizations, researchers could gain insight into how a firm's starting point in terms of environmental performance shapes EMS design and implementation.

In the previous section, we described the ideal conditions for establishing a causal link between EMS adoption and changes in a facility's environmental performance. Under ideal conditions, researchers would randomly assign subjects to one of two groups: the experimental group, which adopts an EMS, and the comparison group, which does not. Today, as firms require their suppliers to adopt EMSs, they are creating a situation that approaches this idealized research experiment. For example, Ford Motor Company and General Motors announced in 1999 that their suppliers must become registered to ISO 14001 over the next several years. Metal finishers that supply these companies must respond to this mandate; metal finishers that supply other industries do not. Assuming that these two groups of metal finishers are otherwise similar (or that any differences could be controlled for), a comparison of their environmental performance over time would lead to a better understanding of whether the mandated adoption of EMSs leads to improved environmental performance. Such research also would help overcome the selection bias issue that arises when studying firms that voluntarily adopt EMSs. Suppliers that respond to customer mandates do not choose to adopt an EMS; they are required to do so in order to stay in business.

The comparative research design that we suggest also could be used to assess policy initiatives, such as performance track programs. Too often policymakers assess these programs in terms of the number of facilities that sign up to participate. Programs with many participants are deemed successful, whereas those with only a handful of participants are viewed more skeptically. We suggest that in addition to the number of participants, programs should be assessed in terms of the degree to which they achieve their stated objectives. If a goal is to improve the environmental performance of firms, then progress toward this goal should be carefully measured and compared with some normal baseline.

In assessing performance track programs, the issue of selection bias will again emerge. Agency programs usually require that firms possess above-average performance to gain entry into a performance track. Facilities that are eligible for EPA's National Environmental Performance Track, for example, must have an established commitment to pollution prevention and a history of regulatory compliance. Participating facilities, by program design, are strong performers. It is likely that if their environmental performance improves after their entry into the performance track, managers are simply continuing an already established commitment to excellence. To assess the ability of performance track programs in promoting change, it will be necessary to compare the environmental performance of participating facilities with the environmental performance of nonparticipating facilities.

Alternatively, the goal of performance track programs might be to change the behavior of agencies, that is, to improve the efficiency of agency compliance assurance programs or to promote a more flexible approach to regulation. If the goal is to change agency behavior, then research should attempt to compare the policies and practices of states where performance track programs have been implemented with those of states that have not undertaken such initiatives. Again, researchers should consider the characteristics that might have caused certain states to adopt performance track programs in the first place, because these characteristics undoubtedly will affect changes in behavior as well.

Conclusion

EMSs are moving forward, and public policy seems to be moving closely behind. Although much remains to be learned about how EMSs may change behavior, many proponents in business and public environmental agencies believe that this approach has promise. They point to improvements in regulatory compliance, innovations in pollution prevention, and reductions in operating costs. Evidence suggests that EMSs will increasingly be used in business-to-business interactions in the future. As more firms with different characteristics adopt this tool, the outcomes associated with EMS implementation may change.

Understanding the role of EMS adoption in environmental performance improvement will become increasingly important as more organizations invest in these systems. We have framed the future of policy debate and policy-relevant research on EMSs. The analysis provided in this book sets forth an important agenda that we hope others will follow.

Index

Note: In page references, n. indicates endnotes, t indicates tables, and f indicates figures.